Journal of Neural Transmission

Supplementum 27

J. A. Obeso, R. Horowski,
and C. D. Marsden (eds.)

Continuous Dopaminergic Stimulation in Parkinson's Disease

Proceedings of the Workshop in Alicante, Spain,
September 22–24, 1986

Springer-Verlag   Wien New York

Dr. J. A. Obeso
Clinica Universitaria, Pamplona, Spain

Dr. R. Horowski
HA Klinische Forschung II, Schering AG, Berlin

Prof. Dr. C. D. Marsden
University Department of Clinical Neurology, The National Hospital for Nervous Diseases, London, U.K.

With 58 Figures

ISSN 0303-6995
ISBN-13: 978-3-211-82034-6   e-ISBN-13: 978-3-7091-8954-2
DOI: 10.1007/978-3-7091-8954-2

Foreword

As early as 1817, in his "An Essay on the Shaking Palsy", James Parkinson mentioned time-dependent changes of behavior that modern diagnostic approaches possibly would term "fluctuations in motoricity". Old as these observations are, they do not favor the assumption that modern long-term antiparkinsonian therapy facilitates or triggers such symptoms. It may be, however, that modern antiparkinsonian therapy prolongs the life expectancy of patients with Parkinson's disease. Therefore the patients render susceptible to a state of denervation topography, at which the loss of the cybernetic coordination in neuron function more frequently triggers motor fluctuations. In fact, the management of these symptoms is difficult and a main problem of today's therapy. Therefore, a symposium related to such aspects merits recognition, especially when it combines basic and clinical aspects. Moreover, basic principles established by biochemical and pharmacological approaches such as receptor function, stimulation, supersensitivity, up- and down-regulation or fluctuation have been tried—in this particular symposium on "Continuous Dopaminergic Stimulation in Parkinson's Disease"—to bring into a therapeutic strategy that allows the management of undesirable motor fluctuations. Maintenance of stable plasma dopa levels and continuous dopaminergic receptor stimulation with lisuride infusion techniques are discussed in extenso in these symposium proceedings. The message of this report is summarized best by the suggestion that these techniques are of special importance to numerous patients suffering from motor fluctuations. In this respect, this symposium gave new impulses for therapeutic strategies applicable to clinical practice.

Würzburg, February 1988   P. Riederer

Acknowledgements

The editors want to thank H. J. Ohnesorge and A. Delafuente and their co-workers from Schering Spain for the organization of the workshop. The skilful assistance of Mrs. H. Haghgou and Springer-Verlag Wien-New York in editing this volume is gratefully acknowledged.

We are particularly obliged to Springer-Verlag Wien-New York for including an extract from the very lively and stimulating discussions. For their share in the discussions and for their contributions we are grateful to all scientists who attended the meeting.

The Editors

List of contributors

Agnoli, A., Department of Neurology, University of Rome "La Sapienza", Rome, Italy.

Aljanati, R., Section of Extrapyramidal Diseases, Institute of Neurology, Montevideo, Uruguay.

Antonini, A., Department of Neurology, University of Rome "La Sapienza", Rome, Italy.

Baronti, F., Department of Neurology, University of Rome "La Sapienza", Rome, Italy.

Bellantuono, P., Department of Neurology, University of Rome "La Sapienza", Rome, Italy.

Bittkau, S., Neurologische Universitätsklinik, Würzburg, Federal Republic of Germany.

Boyce, S., MRC Movement Disorders Research Group, University Department of Neurology and Parkinson's Disease Society Research Centre, Institute of Psychiatry and King's College Hospital Medical School, London, U.K.

Brăvi, D., Department of Neurology, University of Rome "La Sapienza", Rome, Italy.

Brughitta, G., Department of Neurology, University of Rome "La Sapienza", Rome, Italy.

Caamaño, J. L., Section of Extrapyramidal Diseases, Institute of Neurology, Montevideo, Uruguay.

Camps, M., Preclinical Research, Sandoz Ltd., Basle, Switzerland.

Chase, T. N., Experimental Therapeutics Branch, National Institute of Neurological and Communicative Disorders and Stroke, Bethesda, MD, U.S.A.

Chouza, C., Section of Extrapyramidal Diseases, Institute of Neurology, Montevideo, Uruguay.

Cortés, R., Preclinical Research, Sandoz Ltd., Basle, Switzerland.

Critchley, P. H. S., University Department of Neurology, and the Parkinson's Disease Society Research Centre, King's College School of Medicine and Dentistry and Institute of Psychiatry, London, U.K.

Fabbrini, G., Experimental Therapeutics Branch, National Institute of Neurological and Communicative Disorders and Stroke, Bethesda, MD, U.S.A.

Fahn, S., Movement Disorder Division, Department of Neurology, Columbia University, New York, NY, U.S.A.

Fernandez Pardal, M., Department of Neurology, Hospital de Clinicas José de San Martin, University of Buenos Aires, Buenos Aires, Argentina.

García, S., Division of Extrapyramidal Diseases, Department of Neurology, Hospital Francés, Buenos Aires, Argentina.

Gatto, M., Department of Neurology, Hospital de Clinicas José de San Martin, University of Buenos Aires, Buenos Aires, Argentina.

Gebhardt, B., Department of Neurology, University of Würzburg, Würzburg, Federal Republic of Germany.

Gerlach, M., Department of Neurology, University of Würzburg, Würzburg, Federal Republic of Germany.

Gershanik, O. S., Division of Extrapyramidal Diseases, Department of Neurology, Hospital Francés, Buenos Aires, Argentina.

Gessa, G. L., Department of Neurosciences, University of Cagliari, Italy.

Grandas Perez, F., University Department of Neurology, and the Parkinson's Disease Society Research Centre, King's College School of Medicine and Dentistry and Institute of Psychiatry, London, U.K.

Horowski, R., Clinical Research, Schering AG, Berlin and Bergkamen, Federal Republic of Germany.

Jackson-Lewis, V., Movement Disorder Division, Department of Neurology, Columbia University, New York, NY, U.S.A.

Jenner, P., MRC Movement Disorder Research Group, University Department of Neurology, and the Parkinson's Disease Society Research Centre, King's College School of Medicine and Dentistry and Institute of Psychiatry, London, U.K.

Jorge, P., Centro Ramon y Cajal, Madrid, Spain.

Juncos, J. L., Experimental Therapeutics Branch, National Institute of Neurological and Communicative Disorders and Stroke, Bethesda, MD, U.S.A.

Klawans, H. L., Department of Neurological Sciences, Rush-Presbyterian St. Luke's Medical Center, Chicago, IL, U.S.A.

Krause, W., Research Laboratories of Schering AG, Berlin and Bergkamen, Federal Republic of Germany.

Kuhn, W., Department of Neurology, University of Würzburg, Würzburg, Federal Republic of Germany.

Leenders, K. L., MRC Cyclotron Unit, Hammersmith Hospital, London, U.K.

Löschmann, P.-A., Research Laboratories of Schering AG, Berlin and Bergkamen, Federal Republic of Germany.

Luquin, M. R., Movement Disorders Unit, Department of Neurology, Clinica Universitaria, Medical School, University of Navarra, Pamplona, Spain.

Martînez Lage, J. M., Movement Disorders Unit, Department of Neurology, Clinica Universitaria, Medical School, University of Navarra, Pamplona, Spain.

Marsden, C. D., University Department of Clinical Neurology, The National Hospital for Nervous Diseases, Queen Square, London, U.K.

de Medina, O., Section of Extrapyramidal Diseases, Institute of Neurology, Montevideo, Uruguay.

Mena, M. A., Centro Ramon y Cajal, Madrid, Spain.

Micheli, F., Department of Neurology, Hospital de Clinicas José de San Martin, University of Buenos Aires, Buenos Aires, Argentina.

Mouradian, M. M., Experimental Therapeutics Branch, National Institute of Neurological and Communicative Disorders and Stroke, Bethesda, MD, U.S.A.

Nieuweboer, B., Research Laboratories of Schering AG, Berlin and Bergkamen, Federal Republic of Germany.

Nutt, J. G., Department of Neurology, Oregon Health Sciences University, Portland, OR, U.S.A.

Obeso, J. A., Movement Disorders Unit, Department of Neurology, Clinica Universitaria, Medical School, University of Navarra, Pamplona, Spain.

Palacios, J. M., Preclinical Research, Sandoz Ltd., Basle, Switzerland.

Parkes, J. D., University Department of Neurology, and the Parkinson's Disease Society Research Centre, King's College School of Medicine and Dentistry and Institute of Psychiatry, London, U.K.

Perez y Gonzalez, N., Department of Neurology, Hospital de Clinicas José de San Martin, University of Buenos Aires, Buenos Aires, Argentina.

Probst, A., Department of Neuropathology, Institute of Pathology, University of Basle, Basle, Switzerland.

Przuntek, H., Department of Neurology, University of Bochum, St.-Josef-Hospital, Bochum, Federal Republic of Germany.

Quinn, N. P., University Department of Neurology, and the Parkinson's Disease Society Research Centre, King's College School of Medicine and Dentistry and Institute of Psychiatry, London, U.K.

Reiriz, J., Centro Ramon y Cajal, Madrid, Spain.

Rettig, K.-J., Research Laboratories of Schering AG, Berlin and Bergkamen, Federal Republic of Germany.

Riederer, P., Clinical Neurochemistry, Department of Psychiatry, University Clinic, Würzburg, Federal Republic of Germany.

Romero, S., Section of Extrapyramidal Diseases, Institute of Neurology, Montevideo, Uruguay.

Ruggieri, St., Department of Neurological Science, University of Rome "La Sapienza", Rome, Italy.

Scaramelli, A., Section of Extrapyramidal Diseases, Institute of Neurology, Montevideo, Uruguay.

Scipioni, O., Division of Extrapyramidal Diseases, Department of Neurology, Hospital Francés, Buenos Aires, Argentina.

Stahl, S. M., Neuroscience Research Centre, Merck Sharp & Dohme Research Laboratories, Harlow, Essex, and Institute of Psychiatry, University of London, London, U.K.

Stocchi, F., Department of Neurological Science, University of Rome "La Sapienza", Rome, Italy.

Suchy, I., Research Laboratories of Schering AG, Berlin and Bergkamen, Federal Republic of Germany.

Vaamonde, J., Movement Disorders Unit, Department of Neurology, Clinica Universitaria, Medical School, University of Navarra, Pamplona, Spain.

Wachtel, H., Research Laboratories of Schering AG, Berlin and Bergkamen, Federal Republic of Germany.

de Yebenes, J. G., Servicio de Neurologia, Hospital Universitario "S Carlos", Madrid, Spain.

Contents

Continuous dopaminergic stimulation in Parkinson's disease

Complications in treatment

Continuous dopaminergic stimulation experimental studies

Listed in Current Contents

Continuous dopaminergic stimulation in Parkinson's disease

J Neural Transm (1988) [Suppl] 27: 3–10

Pathogenetic studies of motor fluctuations in Parkinson's disease

T. N. Chase, M. M. Mouradian, G. Fabbrini, and **J. L. Juncos**

Experimental Therapeutics Branch, National Institute of Neurological and Communicative Disorders and Stroke, Bethesda, Maryland, U.S.A.

Summary. Pharmacokinetic and pharmacodynamic mechanisms for levodopa have been studied in relation to the pathogenesis of the motor fluctuations which complicate advanced Parkinson's disease. Since levodopa clearance from the general circulation was found to be similar in patients with wearing-off or on-off phenomena and those with a stable response to levodopa, peripheral pharmacokinetic factors are unlikely to be involved. Efficacy half-time for levodopa, on the other hand, was significantly reduced in patients with mainly wearing-off fluctuations in comparison to those manifesting a stable response to oral levodopa; individuals with predominantly on-off phenomenon had an even more extreme reduction in the duration of the antiparkinsonian action of levodopa. Conversion from oral to intravenous levodopa treatment immediately stabilized plasma levodopa levels in both the wearing-off and on-off groups; motor variability also decreased, especially in those with wearing-off phenomenon. During 11 days of continuous intravenous levodopa therapy, additional reductions in motor fluctuations occurred in both groups, but at a significantly faster rate in patients with wearing-off than in those with on-off responses. These results suggest that wearing-off phenomenon may arise as a consequence of the degeneration of dopamine terminals due to natural disease progression with a resultant inability to buffer variations in levodopa availability; on-off phenomenon, may reflect additional postsynaptic dopamine receptor dysregulation, possibly in response to the resultant, nonphysiologic fluctuations in synaptic dopamine.

Introduction

Levodopa initially provides stable symptomatic relief to a majority of those who suffer from Parkinson's disease. Within a few years, however, many of these individuals begin to evidence fluctuations in their antiparkinsonian response (Barbeau, 1974; Marsden and Parkes, 1976; Shaw et al., 1980). Two major patterns are now recognized: wearing-off phenomenon in which motor performance seems to vary in phase with the plasma levodopa oscillations which attend the oral administration of this drug, and the on-off phenomenon in which motor function changes precipitously and in no apparent relation to the timing of levodopa ingestion or other external factors (Marsden and Parkes, 1976).

The pathogenesis of these response fluctuations remains a matter of speculation. Earlier studies suggested that alterations in levodopa absorption, such as might result from the consumption of certain foods, could be responsible (Morgan et al., 1971; Nutt et al., 1984), although more recent investigations have tended to diminish the general importance of this possibility (Juncos et al., 1987 b). Oscillations in plasma levodopa levels have been presumed to be a contributory factor in view of the timing of wearing-off episodes and their attenuation by procedures which stabilize circulating drug concentrations (Shoulson et al., 1975; Nutt et al., 1984; Hardie et al., 1984; Chase et al., 1986). Interference with levodopa transport across the blood brain barrier by other neutral amino acids or by certain products of levodopa metabolism have also been implicated (Reches and Fahn, 1982; Gervas et al., 1983; Nutt et al., 1984); again, however, recent studies cast doubt on the general clinical significance of these possibilities (Juncos et al., 1987 b; Fabbrini et al., 1987 a). A pathogenetic role for the dopamine receptor changes which attend Parkinson's disease and its dopaminomimetic therapy has also been suggested by several previous investigations, although defining the exact nature of these receptor modifications has proven elusive (Lee et al., 1978; Fahn, 1980).

Peripheral pharmacokinetic mechanisms

Altered mechanisms for the elimination of levodopa from the general circulation could potentially explain the appearance of wearing-off or on-off phenomena. Since not all parkinsonian patients receiving chronic levodopa therapy develop these fluctuations, it is conceivable that the ones who do might handle this drug differently. Indeed, a relation between plasma levodopa variations and clinical response fluctuations has been frequently observed (Tolosa et al., 1975; Shoulson et al., 1975) and patients with motor fluctuations reportedly have greater oscillations in plasma no levodopa than those maintaining a stable response (Papavasiliou et al., 1979). Moreover, the metabolic clearance of many drugs change during chronic administration (Cadwallader, 1983). Nevertheless, in a study of patients with idiopathic Parkinson's disease in whom levodopa-carbidopa (Sinemet) was the primary or exclusive antiparkinsonian treatment, neither the clearance of levodopa (calculated by dividing the optimal infusion rate by the mean plasma concentration under steady state conditions) nor its elimination half-life from plasma differed between patients manifesting a stable response to levodopa and those with fluctuations of either the wearing-off or on-off type (Fabbrini et al., 1987 b). These results are consistent with reports that the plasma half-life for levodopa is similar in parkinsonian patients and control subjects (Rossor et al., 1980; Nutt et al., 1984; Nutt et al., 1985).

Central pharmcodynamic mechanisms

The duration of the antiparkinsonian action of levodopa was studied following abrupt cessation of the drug's stable (for at least 16 hours), optimal-dose intravenous infusion (Fabbrini et al., 1987 b). Parkinsonian signs returned at significantly different rates for each of the following levodopa treatment categories: the response half-time (time requires for a 50 percent reduction in the antiparkinsonian efficacy of levodopa) was longest in those who had never before received levodopa or other dopaminomimetic therapy; stable responders, who had no clinically appreciable fluctuations in motor performance while receiving oral levodopa every 4–6 hours, and wearing-off responders, who no evidenced fluctuations cleary related to the timing of oral levodopa-carbidopa when administered every 2–3 hours, had reductions averaging approximately 73 and 82 percent, respectively; the decrease in patients evidencing abrupt shifts between on and off periods in no apparent relation to the timing of levodopa-carbidopa ingestion averaged about 93 percent. For those never before treated with levodopa as well as those with a stable response to this drug, the amount of decline was minimal throughout the 6-hour observation period. One hour after levodopa withdrawal, patients in the wearing-off group had a mild worsening of their parkinsonian scores, in contrast to those with on-off phenomenon who manifested a substantial return of symptoms.

The duration of action of levodopa may provide a quantitative index to the response patterns which occur in parkinsonian patients treated with this drug. In view of the absence of any peripheral pharmacokinetic explanation for the development of motor fluctuations, the observed differences in levodopa efficacy half-time suggest involvement of central pharmacokinetic or pharmacodynamic mechanisms. An increased sensitivity to altered plasma and presumably cerebral levodopa levels is clearly indicated by the rapid clinical deterioration occurring in patients with wearing-off, and especially in those with on-off phenomena.

The progressive loss of dopamine neurons, estimated to exceed 80 percent at onset of parkinsonian symptoms (Bernheimer et al., 1973), may contribute to the observed differences in the duration of the antiparkinsonian action of levodopa. In patients with advanced disease, most cerebral dopamine synthesized from exogenous sources presumably originates form other dopa decarboxylase containing cells having little capacity to store or release the transmitter amine under physiologic conditions (Melamed et al., 1980). Dopamine levels at postsynaptic receptor sites could thus be expected to rapidly reflect the shifts in cerebral levodopa availability which attend the periodic oral administration of the dopamine precursor.

Wearing-off phenomenon may thus arise as a consequence of the progressive loss of dopaminergic neurons, ultimately resulting in a functionally significant reduction in the ability of the dopamine system to buffer fluxes in cerebral levodopa levels (Doller and Connor, 1980; Ogasahara et al., 1984; Juncos et al., 1985). This diminished buffering capacity presumably accounts for the decrease in efficacy half-time as well as for the shortened optimal interdose interval for levodopa. No additional changes need be postulated to explain wearing-off responses, since stabilization of plasma levodopa levels abolishes motor fluctuations of this type (Shoulson et al., 1975; Nutt et al., 1984; Hardie et al., 1984; Chase et al., 1986).

Further insight into the pathophysiology of both wearing-off and on-off responses may derive from an evaluation of the time course of changes in motor fluctuations which follow the initial stabilization of plasma levodopa levels (Mouradian et al., 1987). On the first day of intravenous levodopa therapy, plasma levodopa variability decreased by an average of about 75 percent in both the wearing-off and on-off groups; with continuing intravenous treatment, the degree of plasma levodopa stabilization remained essentially unchanged. Parkinsonian variance, reflecting motor response fluctuations, initially declined by an average of 52 percent in patients with wearing-off phenomenon, but only by 12% in those with on-off. Approximately half of patients judged clinically to have mainly wearing-off phenomenon had at least a 75 percent reduction in response fluctuations. In contrast, despite maintenance of steady plasma levodopa levels, some individuals considered clinically to have predominantly on-off phenomenon continued to evidence some variations in motor performance. Indeed, only about half of those thought to have on-off phenomenon exhibited more than a 25 percent immediate decline in motor fluctuations. Subsequent improvement in response fluctuations proceeded at different rates for patients in the wearing-off and on-off groups: the time required to achieve a 50 percent decline in parkinsonian variance averaged 3.9 days in those judged clinically to have mainly wearing-off phenomenon compared with 8.0 days in those considered to have on-off. After up to 11 days of infusion, parkinsonian variance was reduced by 77 percent in the wearing-off group and by 42 percent in the on-off group (Mouradian et al., 1987).

These results are consistent with previous reports that the stabilization of plasma, and thus presumably of cerebral, levodopa levels tend to diminish motor response fluctuations in parkinsonian patients (Shoulson et al., 1975; Nutt et al., 1984; Hardie et al., 1984; Chase et al., 1986). The finding that the degree of motor stabilization following conversion from periodic oral to continuous intravenous levodopa treatment was substantially greater in the wearing-off than on-off group, despite equivalent reductions in plasma levodopa variations, lends further support to the view that changes in peripheral pharmacokinetic factors do not account for the appearance of motor fluctuations. The rate of stabilization for the wearing-off and on-off groups also differed significantly. About half of patients judged clinically to have primarily wearing-off phenomenon had a rapid and nearly total remission of motor fluctuations; those considered to have mainly on-off phenomenon, on the other hand, exhibited a relatively slow and incomplete decline in their response variations. These differences suggest a basis for a more rigorous separation of these phenomena.

Concluding speculations

Both wearing-off and on-off phenomena have in the past been defined on the basis of the motor response to oral levodopa: as wearing-off phenomenon develop, plasma levodopa oscillations, reflecting the timing of drug ingestion, become increasingly evident clinically; on-off phenomenon, on the other hand, have no obvious relation to identifiable exogenous factors such as diet, physical activity, or dose timing. The present results suggest an alternative definition of these two phenomena: fluctuations of the wearing-off type are those which disappear immediately upon stabilization of plasma levodopa levels, while fluctuations of the on-off type are those which persist over a period of days or weeks. Parkinsonian patients receiving long-term levodopa therapy initially develop fluctuations of the wearing-off type; only later do on-off type responses become clinically evident in some patients. It now appears that a continuum of motor fluctuations exists beginning with pure wearing-off phenomenon, which later may become mixed with increasing proportions of response variations of the on-off type. Ultimately, on-off responses may come to dominate the clinical picture.

The foregoing observations may have pathophysiologic implications. As already discussed, the degeneration of dopaminergic neurons may lead to a proportional decline in the brain's ability to regulate the synthesis, storage and release of dopamine. Consequently, levels of the transmitter amine at postsynaptic receptor sites come to reflect the oscillations in precursor availability which characteristically accompany oral levodopa therapy rather than the relatively stable intrasynaptic concentrations presumed to be normally present in this tonically active system (Bunney et al., 1973). This situation favors the clinical appearance of wearing-off phenomenon. On the other hand, the relatively slower response of on-off phenomenon to continuous parenteral levodopa or to long-acting dopamine agonists (Kartzinel and Calne, 1976), may reflect factors beyond the degeneration of dopaminergic terminals. The gradual amelioration of on-off fluctuations over several days of constant levodopa infusion could indicate the normalization of some neural mechanism, possibly involving the postsynaptic dopamine receptor, which had become dysfunctional as a consequence of chronic, nonphysiologic shifts in stimulation intensity.

The results described above might also have etiologic implications. Most likely the presynaptic dopamine neuron degeneration, which is here suggested to result eventually in the appearance of wearing-off phenomenon, is attributable to natural disease progression. The rapid improvement of this response when plasma levodopa levels are stabilized presumably reflects the functional correction of the brain's diminished capacity to buffer oscillations in amine precursor levels. Support for this view derives from our finding that the degree of this immediate improvement correlates best with disease duration (Mouradian et al., 1987). Postsynaptic dopamine receptor alterations, now suggested as a factor possibly contributing to the appearance of on-off phenomenon, may occur in response to the chronic, intermittent administration of levodopa, which in the absence of usual buffering results in the phasic release of dopamine at receptor sites which normally operate tonically (Bunney et al., 1973; Barbeau, 1974; Lee et al., 1978; Hornykiewicz, 1979). Motor fluctuations of the wearing-off and on-off types may thus be attributable to both natural disease progression and to oral levodopa toxicity, with the latter factor becoming operative only when the former has progressed to a functionally significant degree.

The foregoing considerations could have relevance to the development of improved therapies for the relief of parkinsonian symptoms. In patients with motor fluctuations, the immediate response to therapeutic approaches which result in relatively stable dopaminomimetic levels can be expected to vary with the degree that wearing-off phenomenon is mixed with on-off phenomenon: patients with pure wearing-off should experience rapid relief from their motor fluctuations; individuals disabled by preponderant on-off responses will derive less immediate benefit (Juncos et al., 1987 a). Ultimately, however, all can be expected to improve as a result of constant drug levels. Whether those with mainly on-off phenomenon will be restored to the degree of stability evidenced by those with pure wearing-off remains to be determined. Since the receptor changes which may contribute to the appearance of on-off responses could arise as a result of their chronic, intermittent stimulation, due to the periodic administration of drugs with a relatively short half-life, the long-term consequence of the early and continuing maintenance of constant dopaminomimetic drug levels could be a delay in the onset or a diminution in the severity of on-off responses. This possibility should encourage continuing attempts to develop long half-life or controlled release dopaminomimetics for oral administration (Juncos et al., 1985; Chase et al., 1986; Juncos et al., 1987 a; Chase et al., 1987) as well as practical approaches to the continuous parenteral administration of existing antiparkinsonian preparations (Juncos et al., 1987 c).

References

Barbeau A (1974) The clinical physiology of side effects in long-term L-dopa therapy. Adv Neurol 5: 347–365

Bernheimer H, Birkmayer W, Hornykiewicz O, Jellinger K, Seitelberger F (1973) Brain dopamine and the syndromes of Parkinson and Huntington. Clinical, morphological and neurochemical correlations. J Neurol Sci 20: 415–455

Bunney BS, Walters JR, Roth RH, Aghajanian G (1973) Dopaminergic neurons: effects of antipsychotic drugs and amphetamine on single cell activity. J Pharmacol Exp Ther 185: 560–571

Cadwallader DE (1983) Biopharmaceutics and drug interactions. Raven Press, New York, pp 123–133

Chase TN, Juncos J, Serrati C, Fabbrini G, Bruno G (1986) Fluctuation in response to chronic levodopa therapy: pathogenetic and therapeutic considerations. Adv Neurol 45: 477–480

Chase TN, Juncos JL, Fabbrini G, Mouradian MM (1987) Controlled release levodopa-carbidopa preparations in the treatment of motor response fluctuations in Parkinson's disease. In: Clifford-Rose F (ed) Parkinson's disease. J Libby, London, pp 163–167

Doller HJ, Connor JD (1980) Changes in neostriatal dopamine concentrations in response to levodopa infusion. J Neurochem 34: 1264–1269

Fabbrini G, Juncos JL, Mouradian MM, Serrati C, Chase TN (1987 a) 3-0-methyldopa and motor fluctuations in Parkinson's disease. Neurology 37: 856–859

Fabbrini G, Juncos JL, Mouradian MM, Serrati C, Chase TN (1987 b) Levodopa pharmacokinetic mechanisms and motor fluctuations in Parkinson's disease. Ann Neurol 21: 370–376

Fahn S (1980) "On-off" phenomenon with levodopa therapy in parkinsonism. Neurology 24: 431–441

Gervas JJ, Muradas V, Bazan E, Aguado EG, de Yebenes JG (1983) Effects of 3-OM-dopa on monoamine metabolism in rat brain. Neurology 33: 278–282

Hardie RJ, Lees AJ, Stern GM (1984) On-off fluctuations in Parkinson's disease: a clinical and neuropharmacological study. Brain 107: 487–506

Hornykiewicz O (1979) Compensatory biochemical changes at the striatal dopamine synapse in Parkinson's disease – limitations of L-dopa therapy. Adv Neurol 24: 275–281

Juncos JL, Serrati C, Fabbrini G, Chase TN (1985) Fluctuating levodopa concentrations and Parkinson's disease. Lancet 2: 440

Juncos JL, Fabbrini G, Mouradian MM, Serrati C, Kask AM, Chase TN (1987 a) Controlled release levodopa treatment of motor fluctuations in Parkinson's disease. J Neurol Neurosurg Psychiatry 50: 194–198

Juncos JL, Fabbrini G, Mouradian MM, Serrati C, Chase TN (1987 b) Dietary influences on the antiparkinsonian response to levodopa. Arch Neurol 44: 1003–1005

Juncos JL, Mouradian MM, Fabbrini G, Serrati C, Chase TN (1987 c) Levodopa methyl ester treatment of Parkinson's disease. Neurology 37: 1242–1245

Kartzinel R, Calne DB (1976) Studies with bromocriptine Part 1. "On-off" phenomena. Neurology 26: 508–510

Lee T, Seeman P, Rajput A, Farley IJ, Hornykiewicz O (1978) Receptor basis for dopaminergic supersensitivity in Parkinson's disease. Nature 273: 59–61

Melamed E, Hefti F, Wurtman RJ (1980) Nonaminergic striatal neurons convert exogenous L-dopa to dopamine in parkinsonism. Ann Neurol 8: 558–563

Marsden CD, Parkes JD (1976) "On-off" effects in patients with Parkinson's disease on chronic levodopa therapy. Lancet 1: 292–296

Morgan JP, Bianchine JR, Spiegel HE, Rivera-Calimlin L, Hersey RM (1971) Metabolism of levodopa in patients with Parkinson's disease. Arch Neurol 25: 39–44

Mouradian MM, Juncos JL, Fabbrini G, Chase TN (1987) Motor fluctuations in Parkinson's disease: pathogenetic studies and parenteral levodopa therapy. Ann Neurol 22: 475–479

Nutt JG, Woodward WR, Hammerstad JP, Carter JH, Anderson JL (1984) The "on-off" phenomenon in Parkinson's disease. Relation to the levodopa absorption and transport. N Engl J Med 310: 483–488

Nutt JG, Woodward WR, Anderson JL (1985) The effect of carbidopa on the pharmacokinetics of intravenous administered levodopa: The mechanism of action in the treatment of parkinsonism. Ann Neurol 18: 537–543

Ogasahara S, Nishikawa Y, Yakahashi M, Wade K, Nakamura Y, Yorifuji S, Tarui S (1984) Dopamine metabolism in the central nervous system after discontinuation of L-dopa therapy in patients with Parkinson disease. J Neurol Sci 66: 151–163

Papavasiliou PS, McDowell FH, Wang YY, Rosal U, Miller ST (1979) Plasma dopa and growth hormone in parkinsonism: oscillations in symptoms. Neurology 29: 194–200

Reches A, Fahn S (1982) 3-0-methyldopa blocks dopa metabolism in rat corpus striatum. Ann Neurol 12: 267–271

Rossor MN, Watkins J, Brown MJ, Reid JL, Dollery CT (1980) Plasma levodopa, dopamine and therapeutic response following levodopa therapy of parkinsonian patients. J Neurol Sci 46: 385–392

Shaw KM, Lees AJ, Stern GM (1980) The impact of treatment with levodopa in Parkinson's
    disease. Q J Med 49: 283–293
Shoulson I, Glaubiger GA, Chase TN (1975) On-off response: clinical and biochemical
    correlations during oral and intravenous levodopa administration in parkinsonian pa-
    tients. Neurology 25: 1144–1148
Tolosa ES, Martin WE, Cohen HP, Jacobson RL (1975) Patterns of clinical response and
    plasma dopa levels in Parkinson's disease. Neurology 25: 177–183

Authors' address: Dr. T. N. Chase, National Institutes of Health/NINCDS, Building
10, Room 5C103, 9000 Rockville Pike, Bethesda, MD 20892, U.S.A.

J Neural Transm (1988) [Suppl] 27: 11–15

The case for and concerns about continuous dopamine stimulation in Parkinson's disease*

J. G. Nutt

Department of Neurology, Oregon Health Sciences University, Portland, Oregon, U.S.A.

Summary: Delivery of levodopa or other dopaminergic agents to the striatum is a critical determinant of the clinical response and may account for many of the fluctations in response ("on-off"). This should be manageable with innovative delivery systems for the drugs. However further studies are required to determine if continuous dopaminergic stimulation will be complicated by increased levels of drug metabolites or down regulation of dopaminergic neurotransmission.

Introduction

The chronic treatment of parkinsonian patients with levodopa is complicated by the fluctuating response to levodopa in a large portion of the patients (McDowell et al., 1979). These fluctuations between parkinsonism and mobility, often with dyskinesia, are frequently complicated by their unpredictable nature leading to complete disruption of the patient's daily activities.

The fluctuating response initially appeared to be largely unpredictable with no rational explanation. However, it is becoming increasingly evident that many of the fluctuations are determined by the delivery of levodopa to the brain; that is, the fluctuations are determined by the pharmacokinetics of levodopa. The evidence for this is based on a number of observations. First, Quinn et al. demonstrated that a patient's hourly response to levodopa throughout a single day might appear chaotic but if responses were summed for a number of days it became apparent that "on" periods tended to follow doses of medications (Quinn et al., 1984). Second, Tolosa et al. and Muenter et al. demonstrated that high plasma levodopa concentrations tended to correlate with improved motor function and low plasma concentrations were associated with exacerbation of the parkinsonism (Tolosa et al., 1975; Muenter et al., 1977). Third, slowing or reducing the gastrointestinal absorption of levodopa delayed or reduced the response to the medication (Rivera-Calmlim et al. 1970; Nutt, 1986). Fourth,

---

* Presented in part at the meeting "Continuous Dopaminergic Stimulation", Alicante, Spain, September 1986.

alteration of transport of levodopa from plasma to brain by competition between levodopa and the other large neutral amino acids for the saturable large neutral amino transport system in the brain capillaries altered the response to levodopa (Nutt et al., 1984; Pincus and Barry, 1986). Fifth, the clinical response to short infusions of levodopa was predictable and dose responsive (Nutt and Wood-ward, 1986). Finally, the administration of levodopa by continuous intravenous infusion or duodenal infusion markedly reduced or abolished the fluctuations for hours and perhaps days (Quinn et al., 1984; Nutt et al., 1984; Shoulson et al., 1975; Hardie, 1984; Juncos et al., 1985; Kurlan et al., 1986).

Considering the short plasma half-life of levodopa (Nutt et al., 1985; Hardie et al., 1986), the variable gastrointestinal absorption and the variable transport at the blood-brain barrier, it seems predictable that the clinical response to each single oral dose of medication would often be unpredictable. Perhaps the more pertinent question is not why the "on-off" phenomena develops but why it is not apparent immediately upon initiation of levodopa therapy.

The recognition that delivery of levodopa to the brain is a critical determinant of the clinical response suggests that methods of continuously administering levodopa or other dopaminergic agents may reduce fluctuations in clinical response. There are, however, two immediate concerns regarding this approach. First, the continuous administration of drug generally results in administration of more drug per day than when the drug is given intermittently by oral admin-istration. This may alter the catabolism of the drug, increase the accumulation of metabolites and expose the patient to other cumulative toxic effects of the drug. Secondly, continuous administration of the drug may lead to down-regulation or tolerance to the drug.

Although a variety of metabolites of dopaminergic agents could potentially interfere with the clinical response to the drugs, only 3-0-methyldopa, a me-tabolite of levodopa, has been demonstrated to block levodopa's actions, pre-sumably by competing with levodopa for transport into the brain (Gervas et al., 1983; Reches and Fahn, 1982; Reches et al., 1982; Nutt et al., 1987). As the plasma concentrations of 3-0-methyldopa appear to reflect the daily dose of levodopa (Nutt et al., 1987; Sharpless et al., 1972), the larger daily levodopa dose administered with continuous levodopa infusions or sustained release prep-arations predictably leads to higher plasma concentrations of 3-0-methyldopa. Forty-eight hour levodopa infusions or chronic administration of the sustained-release preparation of levodopa termed CR-3 doubled plasma 3-0-methyldopa (Nutt et al., 1986). However, as argued elsewhere (Nutt et al., 1987), a doubling of 3-0-methyldopa concentrations is probably a relatively small contribution to the total plasma concentration of large neutral amino acids competing with levodopa for transport at the blood-brain barrier and therefore 3-0-methyldopa is unlikely to be a major determinant of levodopa flux into brain.

Dopaminergic down-regulation or tolerance is a second concern with con-tinuous administration. Continuous, but not intermittent, apomorphine or am-phetamine administration leads to a tolerance to the drug effects in rats (Castro et al., 1985; Post, 1980). Is there evidence for down-regulation in the parkinsonian patient chronically treated with levodopa? It is a common clinical observation that many patients treated with levodopa will do relatively well in the morning and then have a declining response to the medication in the afternoon and evening. One explanation for this phenomena would be down-regulation of the dopamine receptor. We have looked for evidence of down-regulation by comparing the response to apomorphine in the morning and afternoon and find no quantitative difference in the response. This observation does not support the hypothesis that the poorer responses in the afternoon represent diurnal down-regulation of the dopamine receptor (Gancher and Nutt, 1987).

The response to continuous levodopa infusions of 24 to 56 hours also raises the question of tolerance. Seven of the 13 patients we have studied had a sustained response for the duration of the infusion. However, on the second day of infusion, 4 patients had a re-emergence of parkinsonism and 2 patients developed increasing dyskinesia despite stable plasma concentrations of levodopa. The interpretation of these clinical responses is complicated in that the studies were conducted to investigate the effects of challenges with various amino acids and furthermore, some of these patients had bromocriptine withdrawn immediately prior to the infusion. Nevertheless these observations raise the question of whether continuous levodopa infusions can produce a continuous clinical response. Even in the patients in whom the response was sustained throughout the infusion, the sensitivity to perturbations by amino acid challenges appeared to be much greater on the second day of the infusion that on the first.

Sustained-release preparations of levodopa produced very high concentrations of plasma levodopa in some patients whose dose was titrated by clinical response (Nutt et al., 1986). However, these patients continued to fluctuate with the "off" periods corresponding to decreased plasma concentrations of levodopa, even though trough concentrations greatly exceeded peak concentrations obtained during administrations of regular Sinemet®. Dopaminergic down-regulation or tolerance is one explanation for this observation.

The transient improvement in response to levodopa after a "drug holiday" has also been interpreted as evidence of down-regulation (Direnfeld et al., 1978; Kay and Feldman, 1986). However, the apparently enhanced response to levodopa after "holidays" has not been rigorously shown to be of greater magnitude or of longer duration and is complicated by a difference in baseline disability before and after the "holiday" (Sweet et al., 1972).

Finally, Cotzias et al. found that the therapeutic response to the chronic administration of the D-1 and D-2 dopamine agonist, N-propylnoraporphine, decreased after several weeks and required repeatedly increasing the dose to maintain the antiparkinsonian action (Cotzias et al., 1976).

Despite these observations suggesting that continuous administration of levodopa may produce down-regulation, other groups have not found any decrement of response when levodopa is administered intravenously for days (Juncos et al., 1985, Chase, this volume). The continuous subcutaneous administration of lisuride, a D-2 agonist, also does not appear to induce down-regulation (Obeso et al., 1986). Resolving these different observations on intermittent and continuous dopaminergic stimulation in animals and humans as well as the role of D-1 and D-2 receptor stimulation in the clinical response are critical issues that will determine the ultimate success of strategies for continuous administration of dopaminergic agents in Parkinson's disease.

Acknowledgements

I thank Dr. S. T. Gancher for helpful comments and Peggy Price for preparation of the manuscript. Supported in part by NIH-NINCDS 2 RO1 NS 21062-04.

References

Castro R, Abreu P, Calzadilla CH, Rodriguez M (1985) Increased or decreased locomotor response in rats following repeated administration of apomorphine depends on dosage interval. Psychopharmacology 85: 333–339

Cotzias GC, Papavasiliou PS, Tolosa ES, Mendez JS, Bell-Midura M (1976) Treatment of Parkinson's disease with apomorphines: Possible role of growth hormone. N Engl J Med 294: 567–572

Direnfeld L, Spero L, Marotta J, Seeman P (1978) The L-Dopa on-off effect in Parkinson's disease: Treatment by transient drug withdrawal and dopamine receptor resensitization. Ann Neurol 4: 573–575

Gancher ST, Nutt JG (1987) Diurnal responsiveness to apomorphine. Neurology 37: 1250–1253

Gervas JJ, Muradas V, Bazan E, Aguado EG, de Yebenes JG (1983) Effects of 3-OM-dopa on monoamine metabolism in rat brain. Neurology 33: 278–282

Hardie RJ, Lees AJ, Stern GM (1984) On-off fluctuations in Parkinson's disease: a clinical and neuropharmacological study. Brain 107: 487–506

Hardie RJ, Malcolm SL, Lees AJ, Stern GM, Allen JG (1986) The pharmacokinetics of intravenous and oral levodopa in patients with Parkinson's disease who exhibit on-off fluctuations. Br J Clin Pharmacol 22: 429–436

Juncos J, Serrati C, Fabbrini G, Chase TN (1985) Fluctuating levodopa concentrations and Parkinson's disease. Lancet 2: 440

Kaye JA, Feldman RG (1986) The role of L-dopa holiday in the long-term management of Parkinson's disease. Clin Neuropharmacol 9: 1–13

Kurlan R, Rubin AJ, Miller C, Rivera-Calmlim L, Clarke A, Shoulson I (1986) Duodenal delivery of levodopa for on-off fluctuations in parkinsonism: preliminary observations. Ann Neurol 20: 262–265

McDowell RH, Papavasiliou P, Sweet R (1979) Long-term study and the effect of human growth hormone in parkinsonian patients treated with L-dopa. Adv Neurol 24: 475–488

Muenter MD, Sharpless NS, Tyce GM, Darley FL (1977) Patterns of dystonia ("I-D-I" and "D-I-D") in response to L-dopa therapy for Parkinson's disease. Mayo Clin Proc 52: 163–174

Nutt JG (1986) Food interactions in parkinsonism. Neurol Consult 3: 1–4

Nutt JG, Woodward WR (1986) Levodopa pharmacokinetics and pharmacodynamics in fluctuating parkinsonian patients. Neurology 36: 739–744

Nutt JG, Woodward WR, Anderson JL (1985) Effect of carbidopa on pharmacokinetics of intravenously administered levodopa: implications for mechanism of action of carbidopa in the treatment of parkinsonism. Ann Neurol 13: 537–543

Nutt JG, Woodward WR, Carter HJ (1986) Clinical and biochemical studies with controlled-release Sinemet. Neurology 36: 1206–1211

Nutt JG, Woodward WR, Gancher ST, Merrick D (1987) 3-0-methyldopa and the response to levodopa in Parkinson's disease. Ann Neurol 21: 584–588

Nutt JG, Woodward WR, Hammerstad JP, Carter JH, Anderson JL (1984) The "on-off" phenomenon in Parkinson's disease: relation to levodopa absorption and transport. N Engl J Med 310: 483–488

Obeso JA, Luquin MR, Martinez-Lage JM (1986) Lisuride infusion pump: a device for the treatment of motor fluctuations in Parkinson's disease. Lancet 1: 467–470

Pincus JH, Barry K (1986) Control of the "on-off" phenomenon with dietary manipulation. Ann Neurol 20: 149

Post RM (1980) Minireview: intermittent versus continuous stimulation: effects of time interval on the development of sensitization or tolerance. Life Sci 26: 1275–1282

Quinn N, Parkes D, Marsden CD (1984) Control of on/off phenomenon by continuous intravenous infusion of levodopa. Neurology 34: 1131–1136

Reches A, Fahn S (1982) 3-0-methyldopa blocks dopa metabolism in rat corpus striatum. Ann Neurol 12: 267–271

Reches A, Mielke LR, Fahn S (1982) 3-0-methyldopa inhibits rotations induced by levodopa in rats after unilateral destruction of the nigrostriatal pathway. Neurology 32: 887–888

Rivera-Calmlim L, Dujovne CA, Morgan JP, Lasagna L, Bianchine JR (1970) L-dopa treatment failure: explanation and correction. Br Med J 4: 93–94

Sharpless NS, Muenter MD, Tyce GM, Owen CA (1972) 3-methoxy-4-hydroxyphenylalanine (3-0-methyldopa) in plasma during oral L-dopa therapy of patients with Parkinson's disease. Clin Chim Acta 37: 359–369

Shoulson I, Glaubiger GA, Chase TN (1975) On-off response: clinical and biochemical correlations during oral and intravenous levodopa administration in parkinsonism patients. Neurology 24: 1144–1148

Sweet RD, Lee JE, Spiegel HE, McDowell F (1972) Enhanced response to low doses of levodopa after withdrawal from chronic treatment. Neurology 22: 520–525

Tolosa ES, Martin WE, Cohen HP, Jacobson RL (1975) Patterns of clinical response and plasma dopa levels in Parkinson's disease. Neurology 25: 177–183

Author's address: Dr. J. G. Nutt, Oregon Health Sciences University, Department of Neurology, 3181 SW Sam Jackson Park Road, Portland, OR 97201, U.S.A.

J Neural Transm (1988) [Suppl] 27: 17–25

Subcutaneous administration of lisuride in the treatment of complex motor fluctuations in Parkinson's disease

J. A. Obeso, M. R. Luquin, J. Vaamonde, and **J. M. Martínez Lage**

Movement Disorders Unit, Department of Neurology, Clínica Universitaria,
Medical School, University of Navarra, Pamplona, Spain

Summary. 28 patients with Parkinson's disease showing complex "on-off" fluctuations in response to chronic levodopa plus dopa decarboxylase inhibitor (po) were treated with subcutaneous lisuride using a portable infusion pump. All patients improved initially during the first weeks of treatment. Four patients abandoned the trial within the first few weeks as a consequence of psychiatric complications (2 cases), inability to understand how to use the pump (one case) and subcutaneous nodule formation plus psychological rejections to wearing a pump (one case). All other 24 patients were treated for a minimum periods of 3 months (mean 9.6 months, maximum 24 months). The average daily dose of lisuride was 2.80 mg. The levodopa dose was reduced by 37%, but total withdrawal was not possible in any patient. Among the 18 patients who continued treatment at present, about 50% are independent and capable of undertaking most daily life activities. Psychiatric side-effects were present in 9 patients leading to permanent withdrawal in five.

Subcutaneous lisuride infusions added to oral levodopa are clearly effective in patients with severe motor fluctuations. Careful selection of suitable patients and close monitoring is mandatory in order to obtain the best therapeutic results while reducing the risk of psychiatric adverse effects.

Introduction

Most patients with Parkinson's disease respond adequately to levodopa during the first years of treatment. In many patients this initial benefit is blurred by the appearance, after 3–5 years of levodopa therapy, of daily fluctuations in motor performance and various types of involuntary movements (Marsden and Parkes, 1976). Therapeutic strategies such as shortening the time interval between each levodopa dose, increasing daily levodopa intake, adding dopamine agonists (i.e. bromocriptine, lisuride, pergolide) or a MAO-B inhibitor (deprenyl), may prove useful in patients with relatively simple fluctuations ("wearing off"). However, such therapeutic strategies generally produce little benefit in patients with more complex fluctuations ("On-Off"). This later type of complication is typically characterized by failure to respond to individual levodopa doses and unpredictable changes in mobility (Marsden et al., 1982). In addition, abnormal involuntary movements and postures may become by themselves an important source of disability.

The demonstration over the past few years that intravenous levodopa or lisuride infusion can correct motor fluctuations in Parkinson's disease (Quinn et al., 1984; Nutt et al., 1984; Hardie et al., 1984; Obeso et al., 1983, 1986 a) led the authors to the idea of developing a portable system for continuous dopaminergic stimulation (Obeso et al., 1986 b). The initial experience has been previously reported and presented in several international meetings (Obeso et al., 1986 c; Obeso 1986; Luquin et al., 1986). In this paper we shall summarize the experience and problems after two years of using this form of treatment.

Patients and method

28 patients (17 females and 11 males) with long-standing Parkinson's disease and severe motor disability associated with chronic levodopa therapy unresponsive to treatment with standard antiparkinsonian drugs have been studied. The general characteristics of these patients and their main motor disability justifying treatment with subcutaneous lisuride are summarized in Tables 1 and 2 respectively. Notable features of this group of patients are the relatively early onset and long duration of the disease, being under levodopa for many years, and suffering severe disability when "off", yet capable of showing almost normal motor performance when "on". These patients therefore represent the prototype of "therapeutically resistant" parkinsonians encountered in any large neurological clinic. All patients were admitted to hospital for adequate evaluation of mobility throughout the day. "On-Off" charts were recorded on several days before and after initiation of treatment. The King's College Hospital Parkinson's disease clinic rating scale was used for motor function assessment during "On" and "Off" periods. Dyskinesias were evaluated in five different grades of severity according to the degree of functional impairment. The degree and severity of night immobility was also rated from grade 0 (normal) to grade V (completely immobile in bed). The authors themselves carried out the clinical evaluations and tests, but patients were taught to self-rate their motor function before leaving hospital.

All patients started to take domperidone 10 mg tds (po) a few days before and during the initial weeks of the infusion to prevent nausea and vomiting. 7 patients receiving bromocriptine and 5 patients under anticholinergic treatment in addition to levodopa plus dopa decarboxylase inhibitor (DDI) stopped both drugs before the study was initiated.

Table 1. General characteristics of patients treated with subcutaneous lisuride infusion

Number: 28 (17 female, 11 male)
Age onset: 46.3 (16–68)
Age when study: 60.1 (28–72)
Duration of illness: 14 years (4–24)
Duration of levodopatherapy: 11 years (4–16)
Daily levodopa dose: 878 ± 93 (400–1,500)
Mean daily number of doses: 6.7 (4–10)
King's College Disability Scale: Off 67.37
                                    On 14.62

Table 2. Main reasons for starting treatment with subcutaneous lisuride infusion
in 28 patients

Complicated "On-Off"-fluctuations and dyskinesias	14
Levodopa resistant "Off" periods + dyskinesias	11
"Benefit of dose" dyskinesias	3
Severe night immobility	24

Oral daily levodopa treatment with levodopa plus DDI was maintained unchanged during the first days of infusion and adjusted subsequently according to the clinical response. Two different portable infusion pumps (Tecensa-62, Madrid, Spain, and Hoechst-insulin pump, Federal Republic of Germany) were used for lisuride administration. Dry lisuride (0.5 to 4.5 mg) was diluted in 3–4 ml of saline for continuous (24 hours) subcutaneous administration. The needle was changed and replaced in the abdomen every 4–5 days initially, but at present most patients do it every 2–3 days. All patients or their close relatives learned to use the pump independently. Initial infusion rates were usually 40 to 60 μg/h increasing according to the clinical response, until a maximum rate of 187.5 μg/h. Accordingly the total daily lisuride dose varied between 0.5 to 4.5 mg, with a mean of 2.61 mg.

Results

Effect on motor function

All but one of the 28 patients included in this report left hospital under treatment with lisuride via subcutaneous administration and oral levodopa plus DDI. The mean hospitalization time once lisuride infusion had been started was 24 days (3-62). All patients showed some improvement in motor function after starting lisuride infusion. However four patients in whom this treatment was interrupted within the first four weeks are not included in the results. In these patients the main reason for abandoning the trial consisted of psychiatric complications in two, inability to use the system at home in one and the developing of local nodules as well as a rejection to permanent dependence upon the pump in the fourth case. In the 24 patients treated for more than 3 months (mean = 9.6 months, maximum 24 months) the average daily dose of lisuride was 2.80 mg (0.5–4.5), generally using a constant infusion rate (mean = 117.3 μg/hour). The levodopa dose was reduced from a mean of 878 mg/day to 558.1/day (37% decrease). Such reduction was particularly striking in two patients with juvenile Parkinson's disease, who required minimal amounts of levodopa (i.e. 25–50 mg once a day), but total and permanent withdrawal of levodopa was not possible in any patient, confirming previous results with intravenous lisuride infusion (Obeso et al., 1986 a).

Most patients showed considerable improvement in motor capacity after the optimum regimen of subcutaneous lisuride and oral levodopa plus DDI was obtained (Table 3). Perhaps the most dramatic effect was that periods of total immobility disappeared. Thus, although motor capacity was not completely stable throughout the day (Fig. 1), patients could be independent most of the time because "Off" periods were considerably less severe (Table 3) (Fig. 1).

20 J. A. Obeso et al.

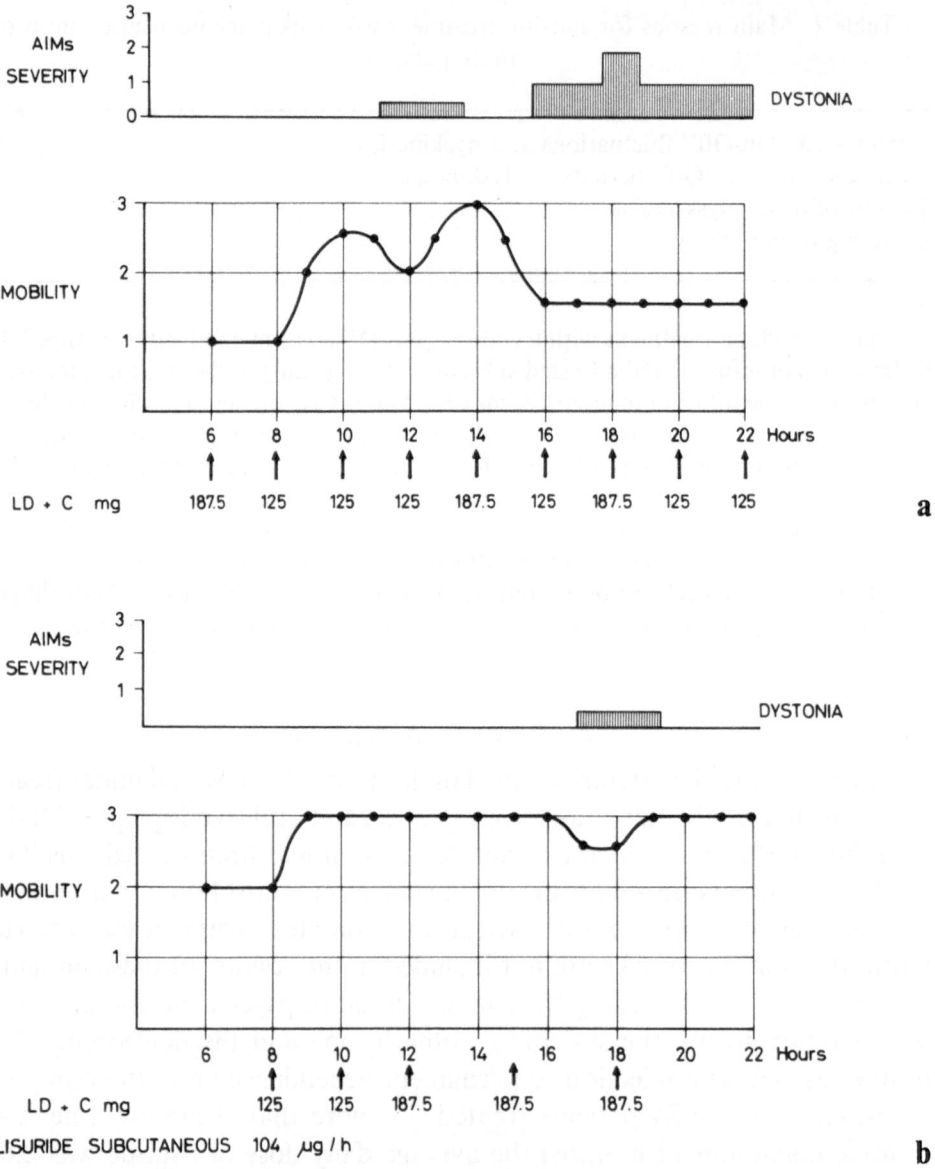

Fig. 1. a Mean disability score (bottom) over the 10 days prior to starting treatment with subcutaneous lisuride. This patients showed a fairly good response during the morning but a poorer one in the afternoon. Periods of immobility were accompanied by severe dystonia (top) mainly affecting the left limbs. **b** Response to subcutaneous lisuride infusion after six months of treatment. Normal mobility was achieved for most of the day, distonia was decreased and "benefit of dose" dyskinesia were not present

Levodopa resistant "Off" periods were virtually abolished so that even lower doses of levodopa could be sufficient to reestablish normal motor performance.

Another notable effect of lisuride infusion was on night and early morning mobility, which were greatly improved in all patients. However, most patients still required their first levodopa dose in the morning to achieve a complete "On".

Table 3. Summary of change in mobility in 24 patients chronically treated with subcutaneous lisuride infusion (X = 9.6 months)

	Control* X		Lisuride* X	
Time "Off"	7.1 hours	(4–12)	0.59	(0–1)
Number "Off" episodes	3.7	(2–5)	0.9	(0–3)
KCH				
"On"	14.62	(12–22)	12.37	(10–8)
"Off"	62.37	(50–92)	25.80	(22–35)
Night mobility	2.7	(1–5)	0.3	(0–2)

* Mean and range

Effect on dyskinesias

In the group of 24 patients, before starting lisuride infusion, 22 had "benefit of dose" choreic movements, 9 showed diphasic dyskinesia and 20 had "Off" period dystonia, the later occurring most often in the morning but usually accompanying any "Off" episode.

Diphasic dyskinesia and "Off" period dystonia, were abolished or greatly reduced in all patients. Patients with "benefit of dose" dyskinesias showed variable response to lisuride infusion. Thus, three patients with "wearing off" fluctuation but severe, generalized chorea as the main cause of disability, obtained a better control of the involuntary movements accompanied by marked attenuation or abolition of the motor oscillations. Among the 14 patients with moderate to severe involuntary movements associated with complex motor fluctuations, the dyskinesias were increased in 5 and unchanged in 9.

Side-effects
Local

Subcutaneous nodules developed at the injection sites in all patients. The tendency to develop nodules varied enormously from one patient to another. Nodules of sufficient dimension and inflammatory component as to interfere with absorption of lisuride (as judged and assumed from clinical observation) were present in 5 patients. In these subjects the nodules appeared within the first 36–48 hours of the injection in any given abdominal site. In 3 patients the nodules became infected on two or three different occasions, requiring treatment with antibiotics (po).

In one patient, the facility for developing nodules as well as physical and psychological inadaptation to the pump led him to discontinue lisuride infusion before returning to hospital.

Psychiatric

Nine patients showed psychiatric complications of sufficient intensity to stop infusion temporarily and responsible for permanent withdrawal in 5 patients. They took the form of visual and/or auditory hallucinations in all 9 patients, organic confusional state was observed in six and a florid paranoid state in five patients. In most subjects these psychiatric manifestations appeared gradually, starting by brief episodes of confusion at night sometime preceded by hallucinations during day-light reaching a state of severe agitation. Eight of the nine patients with psychiatric side-effects were reinitiated on treatment with subcutaneous lisuride infusion, progressively increasing the dose until reaching 50–70% of the previous infusion rate. Three patients had a clear-cut recurrence of psychiatric symptoms within a few days or even in the first hours (one case) of lisuride infusion. Five patients were able to tolerate lisuride subcutaneous infusion again, although three of them have had short periods of visual hallucinations whenever the infusion rate has been slightly increased.

Dyskinesias

In one patient dyskinesias of the type of repetitive alternating movements of the lower limbs, as seen in patients with diphasic dyskinesias, emerged after 4 months of treatment and lasted many hours, being particularly severe (grade IV) in the afternoon. In spite of several changes in the rhythm of infusion and schedule of oral levodopa administrations, it was not possible to improve this situation. Lisuride infusion was thus stopped in this patient.

Benefit of dose choreic dyskinesias have increased or emerged as main motor complications in 5 patients, who required frequent reductions and reorganization of their levodopa regimen.

Gastrointestinal

Nausea and an ill-define feeling of malaise were present in about half the patients at the beginning of treatment. Severe vomiting was not observed in anyone. Most patients stopped taking domperidone after a few months of treatment.

Mechanical

In four patients defective functioning of the pump programs led to delivering a large amount of lisuride (1–1.5 mg) over a short period of time (1–3 hours), provoking a state of deep sleep in all cases and a brief interval of unresponsiveness in one patient. Physical examination, ECG monitoring and routine blood tests did not show any abnormality during such episodes. All patients recovered spontaneously within 4–6 hours showing a 2–3 hours anterograde amnesia.

During the initial months of this experience several patients stopped receiving subcutaneous lisuride for a variable period of 3 to 96 hours, due to defects in the pump mechanisms and in one occasion because of shortage of lisuride supplies. This discontinuation of the infusion did not produce any general or neurological negative effect other than the expected increase in motor disability. At present, most patients stop their pump when taking a bath, swimming or playing some sports, with no harmful consequences.

Biochemical

Laboratory tests looking for liver, kidney or endocrine abnormalities have been consistently normal in all patients. ECG recording has not shown pathological modifications, but Holter monitoring has been done in one patient only. Platelet aggregation was found abnormal in all patients receiving lisuride infusion. This consisted mainly in ADP-induced hypoaggregation, but some patients also showed global hypoaggregation for ADP, collagen and epinephrine. Spontaneous bleeding has not been encountered, but three patients presented small hematomas in the limbs secondary to minor trauma and in the abdomen around the site of the subcutaneous injection.

Evolution

Eighteen of the twenty-four patients included in this report continue under treatment with lisuride infusion at present. Although they still benefit from this form of treatment, the response is not uniform within the group. 5 patients are enjoying normal life. Two have returned to work and another one is particularly noticeable for his present ability to practice sports such a skiing or parachuting. 8 patients are still obtaining a clear benefit from the infusion, but the response is not constant throughout the day. Interestingly most of them show their "semi-off" episodes in the early afternoon. In one of these patients, reduction in motor capacity seems to depend upon limitation of lisuride subcutaneous absorption due to nodule formation. Thus, his disability is minimal during the first day after changing the needle and becomes maximal 48 hours later. Another patient has a very stable, normal, motor function when hospitalized, but presents frequent "off" episodes when submitted to the stress of his job and to social situations which tend to produce irregularities in his levodopa schedule. The last five patients show the most unstable response. This is mainly characterized by a tendency to develop generalized chorea. At present, during prolonged admissions, it is usually possible to regain a certain stability, but this is a brittle pharmacological situation. Most of these patients therefore require frequent readmission and close monitoring. However it must be stressed that these patients are still better, both subjectively and objectively, in their overall clinical situation than before starting on lisuride infusion.

Discussion

These results confirm that continuous dopaminergic stimulation by subcutaneous administration of lisuride can abolish or at least reduce motor fluctuations in patients with severe Parkinson's disease. This effect has been maintained for many months, up to 2 years in one patient, in a significant proportion of subjects. Such findings indicate relative normality of striatal postsynaptic dopaminergic receptors in parkinsonian patients with complex response to standard levodopa therapy. In a number of patients however, the response is not stable, perhaps suggesting certain differences in the biochemical pathology of their diseases or reflecting pharmacodynamic abnormalities induced by prior exposure to other drugs.

Factors so far identified as capable of limiting or conditioning lisuride efficacy are: a) the developing of nodules in the infusion sites, probably interfering with lisuride absorption. b) The interaction with levodopa has probably contributed to maintaining the response for a prolonged period by preventing hyposensitivity, but may also reduce the response of postsynaptic dopaminergic receptors. Thus, in many patients levodopa plus DDI daily doses could be reduced and kept low (Table 2) without compromising motor function. In these subjects, the combination of a continuous (lisuride) stimulus with pulsatile (levodopa oral doses) stimulation may account for the lack of tolerance to lisuride infusions. On the other hand, in 2 patients requiring high levodopa doses (1500 mg and 1800 mg daily), increased levodopa delivery to the brain, 3-OM-dopa accumulations or both factors seemed to reduce lisuride action in the afternoon and evening (Vaamonde et al., 1987). In addition, in patients with severe dyskinesias, these coincided with high levodopa peaks in plasma. c) Psychiatric side-effects were the main reason for infusion withdrawal in the whole group. However, our experience indicates that it may be possible to reinitiate lisuride infusion, albeit at a lower dose, in some cases.

In summary, the experience accumulated over the last two years leads us to conclude tentatively that in treating severe parkinsonian patients with complex fluctuations and dyskinesias, lisuride infusion (plus oral levodopa-DDI) can provide a substantial improvement in motor function. This is almost complete in about 20% of our patients and very significant in 33% of cases, giving rise together to a figure of 50% of patients who have regained an independent life. Such results are not achieved by any other therapeutic maneuvers available at present for clinical application (Fahn, 1982; Quinn, 1984; Melamed, 1986). However, careful selection of suitable patients and close monitoring is mandatory in order to obtain the greatest possible therapeutic benefit while reducing the risk of side effects.

Many questions still need to be answered by future studies. Lisuride infusions may well not be the final pharmacological answer to the treatment of complicated Parkinson's disease. Nevertheless, the knowledge obtained from these and similar studies will probably allow further insight to be obtained into the mechanisms involved in the origin of the complications encountered in the treatment of Parkinson's disease. Above all, these studies may serve to develop more specific dopaminergic drugs and avoid the development of motor and psychiatric complications.

Acknowledgment

The authors are grateful to Schering AG (Berlin and Madrid) for continuous support during this study. This research project was partially funded by the Spanish government (CAYCIT). Mrs. Jacqueline Oehling kindly typed and prepared the manuscript and Mrs. Carlos Elsden reviewed the text.

References

Fahn S (1982) Fluctuations of disability in Parkinson's disease: pathophysiology. In: Marsden CD, Fahn S (eds) Movement disorders. Butterworth, London, pp 123–145

Hardie RJ, Lees AJ, Stern GM (1984) On-off fluctuations in Parkinson's disease. Brain 107: 487–506

Luquin MR, Obeso JA, Martinez, Lage JM, Tresguerres J, Parada J, Nieuweboer B, Dorow R, Horowski R (1986) Parenteral administration of Lisuride in Parkinson's disease. Adv Neurol 45: 561–568

Marsden CD, Parkes JD (1976) "On-off" effects in patients with Parkinson's disease on chronic levodopa therapy. Lancet 1: 292–296

Marsden CD, Parkes JD, Quinn N (1982) Fluctuations of disability in Parkinson's disease: pathophysiological aspects. In: Marsden CD, Fahn S (eds) Movement disorders. Butterworth, London, pp 96–22

Melamed E (1986) Initiation of Levodopa therapy in Parkinsonian patients should be delayed until the advanced stages of the disease. Arch Neurol 43: 402–405

Nutt JG, Woodward WR, Hammerstad JP et al (1984) The "on-off" phenomenon in Parkinson's disease. N Engl J Med 310: 483–488

Obeso JA (1986) Continuous dopaminergic stimulation: lisuride infusions. 4th Eur Workshop Clinical Neuropharmacology. Pamplona

Obeso JA, Martinez Lage JM, Luquin MR (1983) Lisuride infusion for Parkinson's disease. Ann Neurol 14: 134

Obeso JA, Luquin MR, Martinez Lage JM (1986a) Intravenous lisuride corrects motor oscillations in Parkinson's disease. Ann Neurol 19: 31–35

Obeso JA, Luquin MR, Martinez Lage JM (1986b) Lisuride infusion pump: a device for the treatment of motor fluctuations in Parkinson's disease. Lancet 1: 467–470

Obeso JA, Luquin MR, Martinez Lage JM (1986c) Treatment of motor fluctuations in Parkinson's disease by continuous subcutaneous administration of lisuride. Neurology 36 [Suppl 1]: 216

Quinn NP (1984) Anti-parkinsonian drugs today. Drugs 28: 236–262

Quinn NP, Parkes JD, Marsden CD (1984) Control of "on-off" phenomenon by continuous intravenous infusion of levodopa. Neurology 34: 1131–1136

Vaamonde J, Obeso JA, Luqin MR et al (1987) Increased levodopa consumption reduces motor response to apomorphine in Parkinson disease. Neurology 37 [Suppl 1]: 266

Authors' address: Dr. J. A. Obeso, Clínica Universitaria, Apartado 192, E-31080 Pamplona, Spain.

J Neural Transm (1988) [Suppl] 27: 27–33

Subcutaneous lisuride infusion in Parkinson's disease: clinical results using different modes of administration

F. Stocchi, S. Ruggieri, A. Antonini, F. Baronti, G. Brughitta, P. Bellantuono, D. Bravi, and A. Agnoli

Department of Neurology, University of Rome "La Sapienza", Rome, Italy

Summary. The continuous dopaminergic stimulation provided by infusion of dopamine agonist drugs, is a very effective strategy to control ON-OFF fluctuation in Parkinson's disease. Lisuride is a potent dopamine agonist drug, very soluble in water and can be administred subcutaneously. Many authors have shown that the subcutaneous infusion of lisuride can control fluctuations when applied in combination with oral levodopa as a 24 hour continuous infusion regimen. In this study, lisuride was given without any other antiparkinsonian medicament and using a 12 hour infusion regimen wherever possible. 13 fluctuating Parkinsonian patients were studied. 6 out of these 13 were satisfactorily treated with lisuride alone and the remaining 7 with a combination of Lisuride + oral levodopa. Only in 3 out of 13 patients the 24 hour infusion regimen was required.

Introduction

"The on-off phenomenon, in which patients switch rapidly between mobility with dyskinesias (on) and a return of akinesia (off), is unfortunately today an all too familiar problem in the long-term levodopa treatment of Parkinson's disease" (Quinn et al., 1986).

Up to date this problem has not been mastered by using oral levodopa treatment despite a large number of strategies applied. Many studies have shown that in selected patients with severe on-off fluctuations, stable periods of sustained mobility can be achieved during daytime hours and even on consecutive days employing a continuous intravenous infusion regimen of levodopa plus oral peripheral dopa decarboxylase inhibitors (Marion et al., 1986; Nutt et al., 1984; Shoulson et al., 1975). However, a major disadvantage of this therapeutic approach, rendering it unsuitable for everyday's treatment, is the large volume of fluid required. A strategy to avoid this problem is the use of dopamine agonists such as lisuride, which as been shown to be of value in controlling fluctuations in motor performance when given by continuous intravenous infusion (Obeso et al., 1983; Stocchi et al., 1986).

Lisuride also has been administered subcutaneously by an ambulant delivery system to patients with Parkinson's disease and complicated responses to oral therapy. The resultant motor response has been very satisfactory with striking improvement or even prevention of motor fluctuations (Bittkau and Przuntek, 1986; Critchley et al., 1986; Obeso et al., 1986; Ruggieri et al., 1986). In the majority of these studies lisuride has been applied in combination with oral L-dopa as a 24-hour continuous infusion regimen (Bittkau and Przuntek, 1986; Critchley et al., 1986; Obeso et al., 1986).

A major problem of this treatment, as with all agonists, in this particular patient population has been related to psychiatric side effects necessitating discontinuation of the infusion in a number of patients (Critchley et al., 1986). Therefore, in this study the clinical effect of lisuride without oral L-dopa was tested using a 12-hour infusion regimen whenever possible, in order to avoid overdosage more likely to occur during a 24-hour course of application. This treatment regimen also has proved to be highly satisfactory in preventing systemic side effects during prolonged L-dopa infusion therapy (Marion et al., 1986).

Patients and method

13 patients (7 males and 6 females), aged between 32 to 60 years (50.6 ± 9.2), with idiopathic Parkinson's disease (Hoehn and Yahr Rating Scale stage III to V) were studied. All were informed about the study and agreed to participate after full disclosure of its purpose, risks and potential benefits. The duration of symptoms of Parkinson's disease ranged from 4 to 30 years (mean 13 ± 6.57). All patients had been taking a combination of L-dopa (787.3 ± 228.8 mg/day) plus a peripheral dopa decarboxylase inhibitor (DDI) for 4 to 15 years (9.1 ± 3.0; Table 1). The criteria for selection were: Age under 60 years, incapacitating fluctuations in motor performance or very severe abnormal involuntary movements during oral treatment and absence of intellectual impairment or systemic disease. All patients maintained a good response to L-dopa and lisuride i.v. without severe systemic side effects or psychiatric disturbances (Stocchi et al., 1986).

On-off effects have become evident 1–12 years (mean 5.8) after institution of treatment and could be correlated with the timing of L-dopa administration in all patients. 4 patients exhibited drug resistant off periods, and unpredictable on-off fluctuations occurred in 5. On periods were accompanied by dyskinesias in all patients, 6 complained of off-period dystonia and 8 of discomfort due to reduced mobility in bed during the night.

On-off charts, dyskinesia charts (0–4) and the Columbia University Rating Scale (0–100) were used for clinical evaluation.

Lisuride 2 mg (Schering AG, Berlin) was diluted in 2 ml of sterile water and infused subcutaneously by an insulin pump (Lilly cpi 9100; Hoechst MRS I) programmable for varying infusion rates according to the patients' individual daytime requirements. The initial infusion rate was determined on the basis of previous intravenous lisuride infusions in the same patients and adjusted according to therapeutic requirements on the following days.

Subcutaneous lisuride was administered without any additional antiparkinson medication between 8.00 a.m.–8.00 p.m. and discontinued overnight. This regime was maintained for 15 days. In those patients not being completely mobile or still fluctuating under lisuride, oral L-dopa plus DDI was added. A 24-hour infusion regimen was instituted in patients complaining of severe nocturnal akinesia or a delay in turning on in the morning irrespective of daytime infusion. All patients received oral domperidone (60 mg/day) pretreatment to

Table 1. Clinical characteristics of the patients

Patient	Sex	Age	H&Y stage	Years of disease	Years of therapy	Years of LTS	L-Dopa os mg/day	Columbia Off	Columbia On	AIMs	Hours Off
1	f	56	III	12	10	3	750	54	8	2	5
2	m	53	III	4	4	3	750	47	10	1	4
3	m	53	IV	13	12	3	800	60	14	3	5
4	f	55	V	9	7	4	500	63	24	4	3
5	m	37	III	8	7	3	1,000	50	10	2	5
6	f	32	V	16	12	4	750	67	8	2	4
7	m	54	IV	10	10	3	1,000	60	11	4	5
8	f	52	V	14	11	4	500	69	25	4	8
9	m	57	V	10	8	3	1,255	64	19	2	8
10	m	58	III	20	15	3	625	47	11	2	4
11	f	56	IV	8	7	3	1,000	63	21	3	3.5
12	m	36	V	30	6	3	800	61	26	3	3
13	f	60	IV	15	11	6	500	53	15	2	7.5
Mean		50.7		13	9.2	3.3	787.3	58.3	15.5	2.6	5
SD		±9.2		±6.5	±3	±0.8	±228.8	±7.3	±6.6	±1.1	±1.7

prevent systemic side effects. Blood pressure and ECG were monitored during the 24 hours preceding onset of infusion therapy, during the first day and after one week of treatment.

Results

Two groups of patients were identified during the initial 15 days of the study. The first group was formed by 6 patients with constant on periods as soon as the optimal infusion rate of lisuride without additional oral L-dopa medication was established (0.05–0.08 mg/hour; Table 2).

All patients remained mobile throughout each lisuride infusion day and 5 even had a striking relief of their nocturnal akinesia. One patient occasionally exhibited short off periods, specifically such as induced by stressful situations. One patient who continued suffering from nighttime immobility was successfully treated by extending the infusion regimen to 24 hours. However, the night-time infusion rate was diminished (0.065 mg/hour) in relation to the daytime rate (0.08 mg/hour) as the patient had the subjective feeling of his treatment being overdosed and dyskinesias tended to be more severe. Although still present in 5 patients, dyskinesias were less marked in comparison with those under oral treatment. In this group of 6 patients the overall duration of their off periods decreased from 4.3 (3–5) to 0.5 (0–1.5) hours. The delay in turning on in the morning ranged from 30–80 minutes.

In the remaining 7 patients lisuride infusions (rate between 0.06–0.11 mg/hour) had to be supplemented by oral L-dopa (at the minimum effective dose of 200–1000 mg plus DDI; Table 3). These patients displayed prolonged akinetic periods or delays in turning on not improvable by an increased lisuride infusion rate during the preceding 15 days. This therapeutic strategy led to a reduction in the duration of immobility from 5.57 to 0.57 hours on an average. In this group 2 patients out of 7 required a 24-hour regimen since both were severely akinetic in the morning and experienced considerable delays in turning on despite supplemental oral L-dopa. In one of these patients the nocturnal lisuride infusion rate had to be reduced from 0.07 to 0.055 mg/hour.

A comparison of both patient groups revealed that parkinsonians responding to lisuride alone were younger (48 vs. 53 years of age) and had a shorter history of the disease (10.3 vs. 15.3 years), although these differences were not statistically significant (Table 4). Significant differences in the severity of the disease between the two groups did not emanate from the clinical assessment. Likewise, patterns of daily L-dopa intake prior to the infusion regimen and the clinical status under long-term L-dopa treatment were comparable in both groups.

Within the group treated with lisuride plus L-dopa the 2 patients requiring a 24-hour regimen were the most severely affected. In contrast, the severity of the disease assessed in a single patient demanding a 24-hour infusion regimen of lisuride without L-dopa was comparable to that of the other members of his group.

10 patients having been on lisuride infusion for 3–15 months are still under treatment, 4 are on lisuride without the addition of L-dopa, and 6 are receiving

Table 2. Effect of continuous subcutaneous lisuride infusion

Patient	Lisuride mg/hours	Columbia		AIMs	Off status (hours)		Months of treatment	Infusion regimen (hours)	
		Off	On		Infusion	Control			
1	0.065	54	8	1	0	5	15	12	
2	0.06	57	10	1	0	4	12	12	
3	0.06	50	13	2	1	5	3	12	*drop out*
4	0.055	63	22	3	1	3	1	12	*drop out*
6	0.065	36	8	0	0	4	9	12	
5	0.08	35	11	1	1.5	5	11	24	
Mean	0.064	49.1	12	1.3	0.6	4.3	8.5		
SD	±0.008	±11.4	±5.2	±0.6	±0.6	±0.8	±5.4		

Table 3. Effect of continuous subcutaneous lisuride infusion added to oral L-Dopa

Patient	L-Dopa os (mg/day)		Lisuride mg/hour	Columbia		AIMs	Off status (hours)		Months of treatment	Infusion regimen (hours)	
	After 1 month	After 3 months		Off	On		Infusion	Control			
7	625	500	0.11	50	12	3	0	5	10	12	
10	500	375	0.06	30	12	2	1	4	5	12	
11	500	500	0.12	50	21	2	0.5	3.5	5	12	
12	375	375	0.05	61	26	2	0	3	2	12	
13	500	375	0.07	47	15	2	1	7.5	3	12	
8	200	100	0.07	49	24	2	0	8	9	24	*drop out*
9	1,000	875	0.09	50	20	1	0.5	8	7	24	
Mean	528.5	442.8	0.08	48	18.5	2	0.42	5.5	5.8		
SD	±246.8	±232.6	±0.02	±9.2	±5.6	±0.4	±0.44	±2.2	±2.2		

Table 4. Comparison between the two groups

Number of patients	Age	Years of disease	Years of therapy	Columbia		AIMs	Hours Off
				Off	On		
6	47∓10.3	10.3∓4.5	8.6∓3.1	56.1∓10.3	12∓5.2	2.3∓1.1	4.3∓0.9
7	53∓9.7	15.3∓9.4	9.7∓3.4	59.2∓8.3	18∓5.6	2.8∓0.8	5.6∓1.5

lisuride plus oral L-dopa. The mean oral L-dopa intake in these 6 patients decreased from 528.5 ± 246.8 mg/day to 442.8 ± 232.6 mg/day.

All patients experienced side effects such as somnolence, nausea and occasionally vomiting during the first week of treatment despite taking oral domperidone as a pretreatment. They had red nodules at the site of injection disappearing spontaneously within 4–5 days. In one subject (no. 4) the trial was discontinued due to emesis. One patient (no. 8) being on the 24-hour infusion plus oral L-dopa (100 mg) experienced visual hallucinations after 5 months of treatment. These symptoms disappeared gradually when the infusion regimen was reduced to 12 hours. However, the trial had to be discontinued 15 days later while the patient developed paranoia. Patient no. 3 dropped out from the study for reason not concerning the treatment.

Discussion

Lisuride infused subcutaneously improves the symptoms and controls fluctuations of motor performance in complicated Parkinson's disease. These results demonstrate that lisuride provides an effective treatment even without the addition of oral L-dopa. The therapeutic relief has been maintained for several months without any loss of efficacy. On the contrary, oral L-dopa intake was considerably reduced in the group receiving a combined regimen. Patients benefiting from lisuride without additional L-dopa tend to be younger and have a shorter history of the disease whereas the severity of disease has not shown to be a criterion.

The 12-hour regimen has proved to be very effective in most of the patients and possibly prevents psychiatric side effects by avoiding drug accumulation during the night. Evidence is provided by the fact that the only patient suffering from psychiatric side effects was on a 24-hour infusion regimen. Furthermore, the nocturnal infusion rate had to be decreased in other patients on account of severe dyskinesias during the night. However, the 24-hour regimen can be demanding, especially in most severely affected subjects, in which this approach might be regarded as the "ultima ratio".

References

Bittkau S, Przuntek H (1986) Lisuride infusion pump for Parkinson's disease. Lancet 2: 349

Critchley P, Perez FG, Quinn N, Coleman R, Parkes D, Marsden CD (1986) Psychosis and the lisuride pump. Lancet 2: 349

Marion MH, Stocchi F, Quinn NP, Jenner P, Marsden CD (1986) Repeated levodopa infusions in fluctuating Parkinson's disease: clinical and pharmacokinetic data. Clin Neuropharmacol 9: 165–181

Nutt JG, Woodward WR, Hammerstad JP, Carter JH, Anderson JL (1984) The on-off phenomenon in Parkinson's disease: relation to levodopa absorption and transport. N Engl J Med 23: 483–488

Obeso JA, Luquin MR, Martinez-Lage JM (1983) Lisuride infusion for Parkinson's disease. Ann Neurol 14: 134–139

Obeso JA, Luquin MR, Martinez-Lage JM (1986) Lisuride infusion pump: a device for the treatment of motor fluctuations in Parkinson's disease. Lancet 1: 467–470

Quinn N, Marion MH, Stocchi F, Jenner P, Marsden CD (1986) Intravenous dopamine agonist studies in Parkinson's disease. In: Fahn S, Marsden CD, Jenner P, Teychenne P (eds) Recent developments in Parkinson's disease. Raven Press, New York, pp 247–253

Ruggieri S, Stocchi F, Agnoli A (1986) Lisuride infusion pump for Parkinson's disease. Lancet 2: 348–349

Shoulson I, Glaubiger GA, Chase TN (1975) On-off response: clinical and biochemical correlations during oral and intravenous levodopa administration in parkinsonian patients. Neurology 25: 1144–1148

Stocchi F, Ruggieri S, Brughitta G, Agnoli A (1986) Problems in daily performances in Parkinson's disease: the continuous dopaminergic stimulation. J Neural Transm [Suppl] 22: 223–229

Authors' address: Dr. F. Stocchi, 1a Clinica Neurologica d'ell Universita, Viale d'ell Universita 30, I-00185 Rome, Italy.

J Neural Transm (1988) [Suppl] 27: 35–54
© by Springer-Verlag 1988

Chronic s.c. Lisuride in Parkinson's disease — motorperformance and avoidance of psychiatric side effects

S. Bittkau[1] and **H. Przuntek**[2]

[1] Neurologische Universitätsklinik, Würzburg, and [2] Neurologische Universitätsklinik im St. Josef-Hospital, Bochum, Federal Republic of Germany

Summary. On-off fluctuations in longstanding Parkinson's disease initially respond well to a combined drug regime of Levodopa with direct dopamine agonists and L-deprenyl. L-Dopa infusions are efficient, but not applicable for longer use. S.c.-Lisuride-infusions reduce markedly motor-response fluctuations, dystonias and hyperkinesias, but bear the risk of inducing confusion or even psychosis. In patients with coexisting response fluctuations and psychiatric disturbances a therapeutic approach is outlined to preserve still some favourable effects on motor performance avoiding severe psychosis. Side-effects and possible complications of that therapy are discussed as are some further indications for the clinical use of Lisuride in akinetic crisis, the neuroleptic malignant syndrome and in dyskinesias.

Introduction

Motor-response fluctuations are well known complications in longstanding Parkinson's disease (PD): wearing-off phenomena or peak dose hyperkinesias can be correlated with the time of L-Dopa intake, whereas so called "on-off fluctuations" are of higher complexity (Fahn, 1974; Tolosa et al., 1975; Marsden and Parkes, 1976 b; Nutt et al., 1984). Primarily these response fluctuations were thought to be independent of L-Dopa intake, but measurements of L-Dopa plasma-levels as well as of its metabolites showed a certain correlation between response-fluctuations and plasma-level changes (Sweet and Mc Dowell, 1974; Calne et al., 1974; Rivera-Calmlin et al., 1977; Eriksson et al., 1984; Hardie et al., 1985).

So pharmacokinetics of anti-parkinsonian drugs are very important in the treatment of motor response fluctuations, and many attempts to achieve a more constant dopaminergic stimulation have been made:

— L-Dopa infusions (as presented here by Dr. Chase) are not widely applicable for long-term treatment for impracticability and for progressive loss of efficacy (Shoulson et al., 1975; Quinn et al., 1984; Nutt and Woodward, 1986);

— slow release L-Dopa formulations are under clinical investigation and offer

some favourable therapeutic effect (Eckstein et al., 1973; Curzon et al., 1973; Poewe et al., 1986; for further studies, see Marsden et al., 1986, in press);
— the combined use of the MAO-B-inhibitor L-Deprenyl with Levodopa reduces metabolisation of dopamine and thereby increases its bioavailability (Birkmayer et al., 1977; Rinne et al., 1978; Lander et al., 1979; Wajsbort et al., 1982);
— the combined therapy of L-Dopa with dopamine-agonists as Bromocriptine or Lisuride allows by direct dopaminergic stimulation a reduction of L-Dopa intake, it improves markedly motor-performance and there is a definite reduction of fluctuations in motor-performance (Agid et al., 1979; Parkes, 1979; Calne, 1982; Rinne, 1983).

The use of these above-named drugs is mainly limited by possible psychiatric side-effects as confusions, delusions, hallucinations or overt psychoses. All these psychiatric disturbances seem to occur more frequently in the later course of PD and coincide there with the elevated occurrence of motor-response fluctuations (Birkmayer and Riederer, 1975; Hoehn, 1985; Moskowitz et al., 1973; Rinne, 1983; Teychenne et al., 1985).

The therapeutic effect of the constant Lisuride-infusion-pump in diminishing on-off fluctuations first was reported by Obeso et al. (1986). As Lisuride is a strong dopamine-agonist we could see at least confusional episodes in any patient, whom we treated with higher doses of this drug. It was rather soon that the potency of Lisuride to induce psychoses became the most prominent clinical problem in the management of patients with the Lisuride-infusion-pump (Bittkau and Przuntek, 1986; Critchley et al., 1986; Todes, 1986). We therefore selected 5 patients from the 10 Parkinsonian patients we had treated with the Lisuride-pump for motor-response fluctuations. These patients all had a history of previous psychiatric disturbances which in some cases even made hospital care necessary to treat these conditions. In these patients the "therapeutic window" of a still satisfying motor effect with at best no or only little confusion is extremely narrow.

So we concentrate here on the common features in the adjustment of Lisuride-therapy in such patients and we want to demonstrate, how even in these patients a rather good therapeutic effect on motor-response could be achieved.

Patients and methods

Mean-age was 59.2 ± 5.9 yrs, mean duration of illness was 13.4 ± 3.7 years (range 11–20 years). 2 patients were Hoehn and Yahr stage 4 and 3 patients fulfilled the criteria for stage 5.

L-Dopa-treatment had been established for 10.8 ± 3.6 yrs with a mean-intake of 1025 ± 495 mg L-Dopa. The mean off-periods were generally more than 50% of daytime. Further details as relevant medical history, previous episodes of psychotic events and characteristics of motor-response-fluctuations are listed in Table 1. One patient was known of having had two severe psychotic events within 2 years prior to admission (Pat. H. M.), the remaining patients all had developed more than 3 episodes of severe psychosis, which required hospital care. The response fluctuations were "on-off" in 4 patients and complicated end-of-dose deterioration in 1 patient.

Table 1. Characteristics of 5 patients with longstanding PD and response-fluctuations

Patients	C.S.	L.N.	F.L.	K.J.	H.M.	Mean value ± S.D.
Age (yrs)	55	66	57	65	53	59.2 ± 5.9
Sex	female	male	male	male	male	—
Duration of illness (yrs)	12	11	20	12	12	13.4 ± 3.7
Hoehn and Yahr, stage	5	4	5	5	4	4.6 ± 0.5
Etiology of Parkinsonism	idio-pathic	idio-pathic	idio-pathic	toxic (?)	idio-pathic	—
Duration of L-DOPA-treatment (yrs)	10	10	16	6	12	10.8 ± 3.6
Last medication (mg)						
L-DOPA	1,250	625	375	1,500	1,375	1,025 ± 495
Bromocriptine	not tolerated	not tolerated	15	not tolerated	30	—
L-Deprenyl	not tolerated	not tolerated	—	not tolerated	10	—
Budipine	not tolerated	not tolerated	—	not tolerated	80	—
Anticholinerg.	not tolerated	not tolerated	—	not tolerated	—	—
Concomittant diagnosis	stereotactic op.	card. dysrhythmias 2 × M.I.	hypotension			—
Medical history	severe depression hypotension	hypotension	dementia	dementia	—	—
Previous episodes of psychotic events	≥ 3 x, severe	frequent severe	≥ 3	frequent	2 x, severe	—
Response fluctuation	on-off	complicated end of dose	on-off	on-off	on-off	—
Dyskinesias	immobile dystonia (severe)	immobile dystonia	immobile dystonia (severe)	mobile dystonia/ hyperkinesas	severe hyperkinesas	—

Technique of infusion device and dose adaptation: we used the disetronic s.c. infusion pump, the infusion profile is programmable for each hour of the day distinctively. The 3,15 ml reservoir is filled with the appropriate solution of Lisuride in 0.9% NaCl, a very thin catheter is placed under the skin of the thorax or abdomen.

The handling of the system could in all cases be done by relatives of the patient.

After dismissal from hospital patients were seen initially about every two weeks, later in 4–8 weeks intervals. All patients kept their mobility-diaries, to note their fluctuations of mobility as well as tremor or hyperkinesias. Special interest was laid on psychiatric side-effects and a corresponding questioning was constantly carried out when the patient was seen. The aim of therapy was to achieve a good mobility with only some tolerable confusion, but to avoid as much as possible hallucinations or illusions.

The grading of a patient for mobility as well as clinical evaluation by the Columbia University Rating Scale (C.U.R.S., 25 criteria, 0–4 points each, range 0–100 points max.), the King's College Hospital Parkinson's Disease Rating-Scale (King's P.D.R.S., 39 criteria, 0–3 points each, range 0–117 points max.) distinctively for On- and Off-periods and the Zung's depression score (ZUNG, 20 criteria, 1–4 points each, range from 20–80 points max.) was done before initiation of Lisuride treatment ("t_1") and 6 weeks after initiation of therapy (pat. H. M.: t_2 = after 3 weeks) as well as 12 weeks after initiation of therapy (pat. H. M.: t_3 = after 8 weeks). Mobility was rated by the patient or by a patient's relative for total hours of good mobility ("On"), reduced, but preserved mobility ("Intermediate") and of bad or absent mobility ("Off").

Statistical analysis was done by Student's t-test for comparison of the mean values (L-Dopa, evaluation-scores) at time 1 with time 2 *or* time 3, variance-analysis for all values at times 1, 2, *and* 3 was done by the Friedman test, p-values are listed and $p \leqslant 0.05$ is marked with one asterisk (*).

Results

Table 2 gives the results for each individual patient: L-Dopa or Lisuride intake, mobility in hours of on- and intermediate-stages and relevant evaluation scores as well as the mean-values of all named parameters including the respective standard-deviations.

There was a significant decrease of L-Dopa intake (Fig. 1) from 1025 ± 495 mg to 262.2 ± 214 mg (Friedman-test, t_1 t_3, p = 0.021, *), the mean Lisuride-dosis at time 3 was 0.94 ± 0.65 mg/d.

The time in hours expressed for on- and intermediate-stages (Fig. 2) increased significantly from 11.04 ± 5.4 hrs/d to 18.6 ± 4.8 hrs/d (t-test, t_1 t_3, p = 0.012, *, Friedman-test, t_1 t_3, p = 0.025, *).

The mean score of the C.U.R.S. (Fig. 3) decreased significantly from 61.8 ± 19.1 to 35.0 ± 21.9 points, the mean scores of the King's Hospital P.D.R.S. decreased for On-periods from 52.4 ± 19.7 points to 27.6 ± 20.9 points (Fig. 4) this was highly significant by t-test (t_1 t_3, p = 0.007, **) and significant by the Friedman-test (t_1 t_3, p = 0.016, *). The same score for Off-periods decreased from 88.6 ± 10.2 points to 64.8 ± 23.2 points and did not reach significance.

The score by ZUNG as an easy test for depression and the psychic state of the patient decreased significantly from 50.68 ± 9.8 point to 38.6 ± 7.6 point (t-test, t_1 t_3, p = 0.037, *), (Fig. 5).

Table 2. Clinical outcome of chronic s.c. Lisuride in PD

Patient	t_x	Time	L-Dopa	Li-suride	Mobile hrs. % of daytime	CURS	King's PDRS On	King's PDRS Off	ZUNG's
C.S.	t_1	pre	1,250	—	10.0 ± 6.2	87	57	93	62
	t_2	6 wks	562	2.1	79.7 ± 10.4	40	31	78	36
	t_3	12 wks	625	0.9	47.5 ± 35.4	51	27	73	37
L.N.	t_1	pre	625	—	44.6 ± 17.9	53	45	93	56
	t_2	6 wks	312	0.6	83.3 ± 22.1	32	34	75	56
	t_3	38 wks	562	0.8 x	100 ± 1.7	27	5	37	46
F.L.	t_1	pre	375	—	47.5 ± 19.2	71	71	89	52
	t_2	6 wks	187	0.5	68.7 ± 25.0	57	69	88	46
	t_3	12 wks	187	0.2 x	77.1 ± 29.2	63	46	91	47
K.J.	t_1	pre	1,500	—	58.3 ± 32.1	36	67	97	36
	t_2	6 wks	437	2.0	98.3 ± 2.5	19	47	73	30
	t_3	12 wks	187	0.8 x	72.9 ± 9.2	26	51	79	32
H.M.	t_1	pre	1,375	—	69.6 ± 25.0	62	22	71	47
	t_2	4 wks	187	1.0	86.2 ± 13.3	20	15	57	36
	t_3	8 wks	250	2.0	90.0 ± 12.5	8	9	44	31
All patients	t_1		$1,025 \pm 495$	—	46.0 ± 20.1	61.8 ± 19.1	52.4 ± 19.7	88.6 ± 10.2	50.6 ± 9.8
	t_2		337 ± 163	1.37 ± 0.75	83.1 ± 14.8	37.0 ± 15.9	39.2 ± 20.2	74.2 ± 11.2	42.0 ± 11.4
	t_3		362 ± 219	0.94 ± 0.65	77.5 ± 15.2	35.0 ± 21.9	27.6 ± 20.9	64.8 ± 23.2	38.6 ± 7.6
Difference (%) t_1–t_3			−64.7		+68.5	−43.4	−47.3	−26.9	−23.7
T-test (p) t_1 t_2			0.025*	—	0.016*	0.014*	0.039*	0.019*	0.090†
t_1 t_3			0.055†	—	0.012*	0.035*	0.007 **	0.064†	0.037 ˣ
Analysis of variance			0.021*	—	0.025*	0.025*	ˣ	0.099†	0.052†

ˣ Pulsatile
* p ⩽ 0.05
** p ⩽ 0.01
† Not significant

Dose adaptation of Lisuride – avoidance of psychiatric side-effects

In some patients we started with rather high initial doses of Lisuride, e.g. 1.5 mg–2.1 mg/d. For psychiatric side-effects dose reduction was nearly always necessary, the first reduction within 3–10 days after initiation of therapy, and even

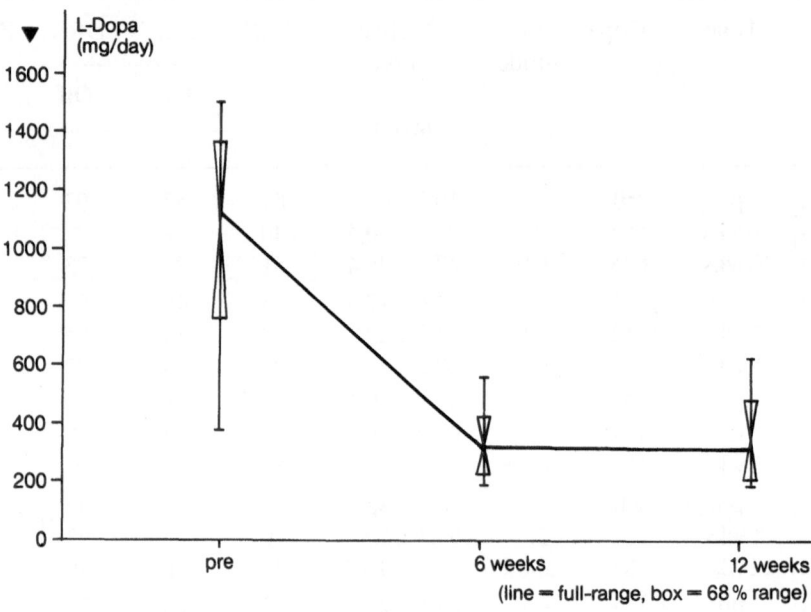

Fig. 1. Mean-L-Dopa intake before (t_1) and after 6 (t_2) and 12 weeks (t_3) of s.c. Lisuride infusion

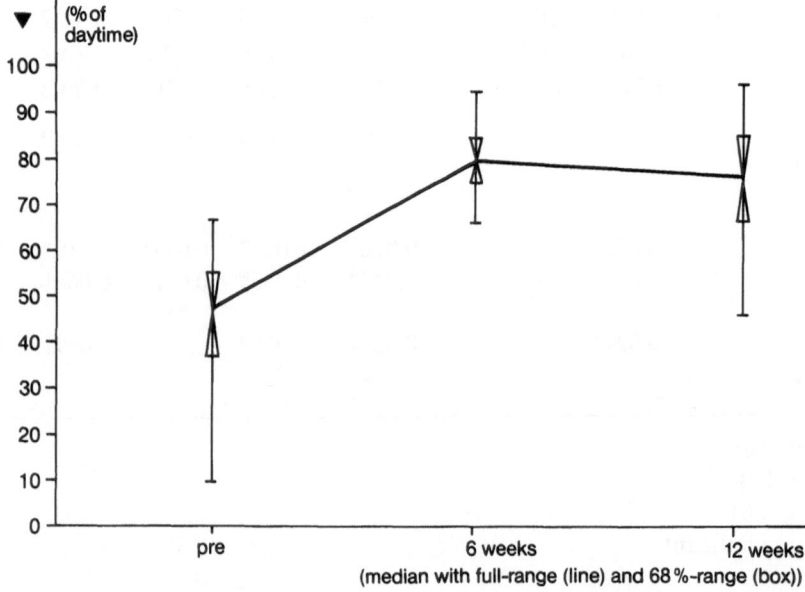

Fig. 2. Increase of preserved motility after addition of s.c. Lisuride to Levodopa (hours in % of daytime of "On" and "Intermediate" phases)

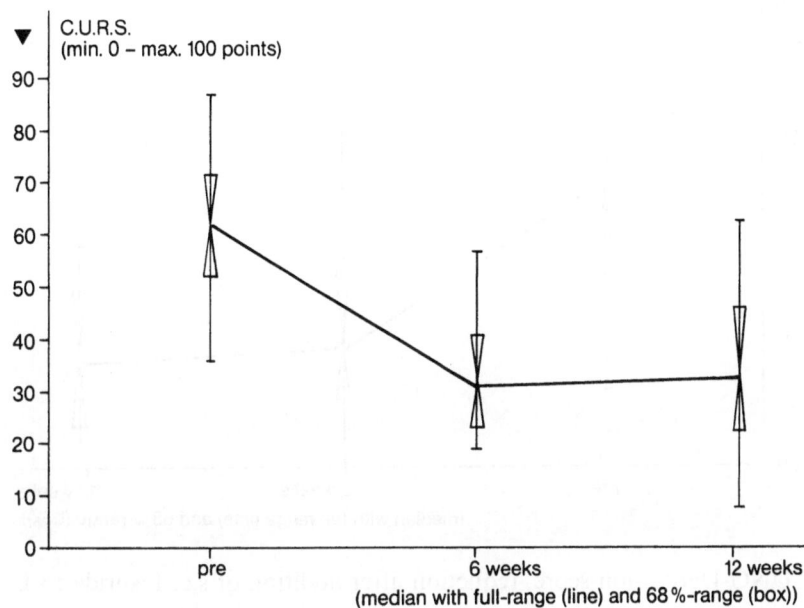

Fig. 3. C.U.R.S. (Columbia University, Rating-Scale) scores before and after addition of s.c. Lisuride to Levodopa

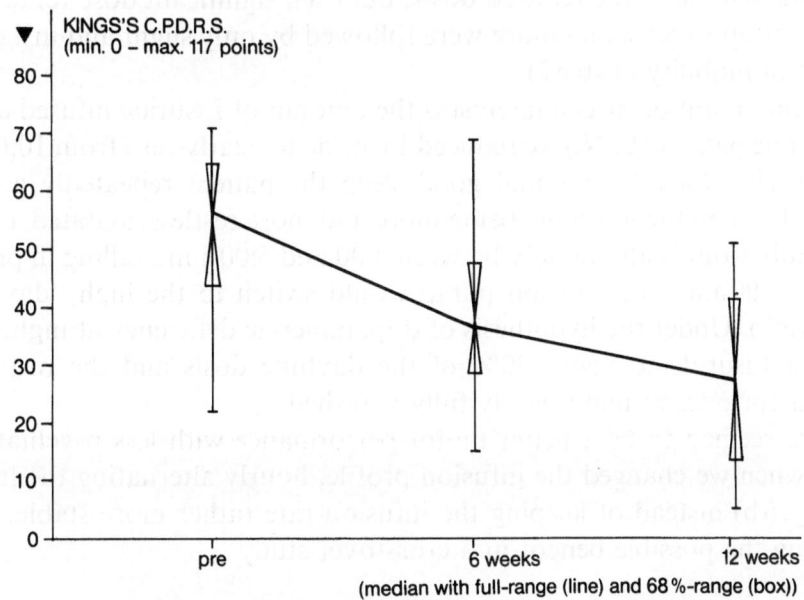

Fig. 4. King's College Parkinson's Disease Rating-Scale: score reduction for On-periods before and after addition of s.c. Lisuride to Levodopa

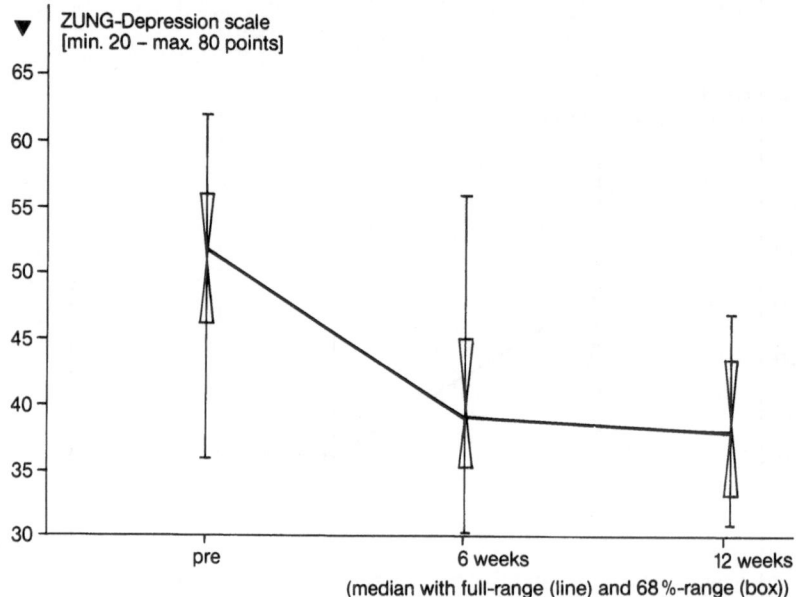

Fig. 5. ZUNG-Depression-score, reduction after addition of s.c. Lisuride to Levodopa

after temporary well-being of the patients for some weeks consistently in all patients further dose-reductions after 6–12 weeks, sometimes in several steps, were necessary as a consequence of developing psychiatric side effects. The optimum motor effect under the maximal Lisuride dosis would not in all cases be maintained under the reduced dosis, but even significant dose reductions of Lisuride of up to 50% and more were followed by only slight and insignificant decrease in mobility (Table 2).

Another point of special interest is the amount of Lisuride infused at night-time. In one patient (L. N.) we reduced Lisuride to nearly zero from 10:00 p.m.–4:00 a.m. (Fig. 6 a). After initial good sleep the patient repeatedly got up at 1:00–2:00 a.m. in the morning, being more and more restless, agitated, confused and usually would calm rapidly between 4:00 and 5:00 a.m., falling in profound sleep (at 4:00 a.m. the infusion pump would switch to the high "daytime-in-fusionrate"). Under the hypothesis of dopaminergic deficiency at nighttime we increased Lisuride to about 20% of the daytime dose and the restless and confused episodes at night nearly fully vanished.

There seemed to be a better motor-performance with less psychiatric side effects, when we changed the infusion profile, hourly alternating the infusion-rate (Fig. 6 b) instead of keeping the infusion-rate rather more stable. We are examining the possible benefit in a cross-over study.

Side effects

Abrupt stop of constant Lisuride infusion, and even more, when in combination with neuroleptics, may lead to a "dopamine-deficiency-syndrome", which even may present with conscience-alterations and coma: one patient (C. S.), generally

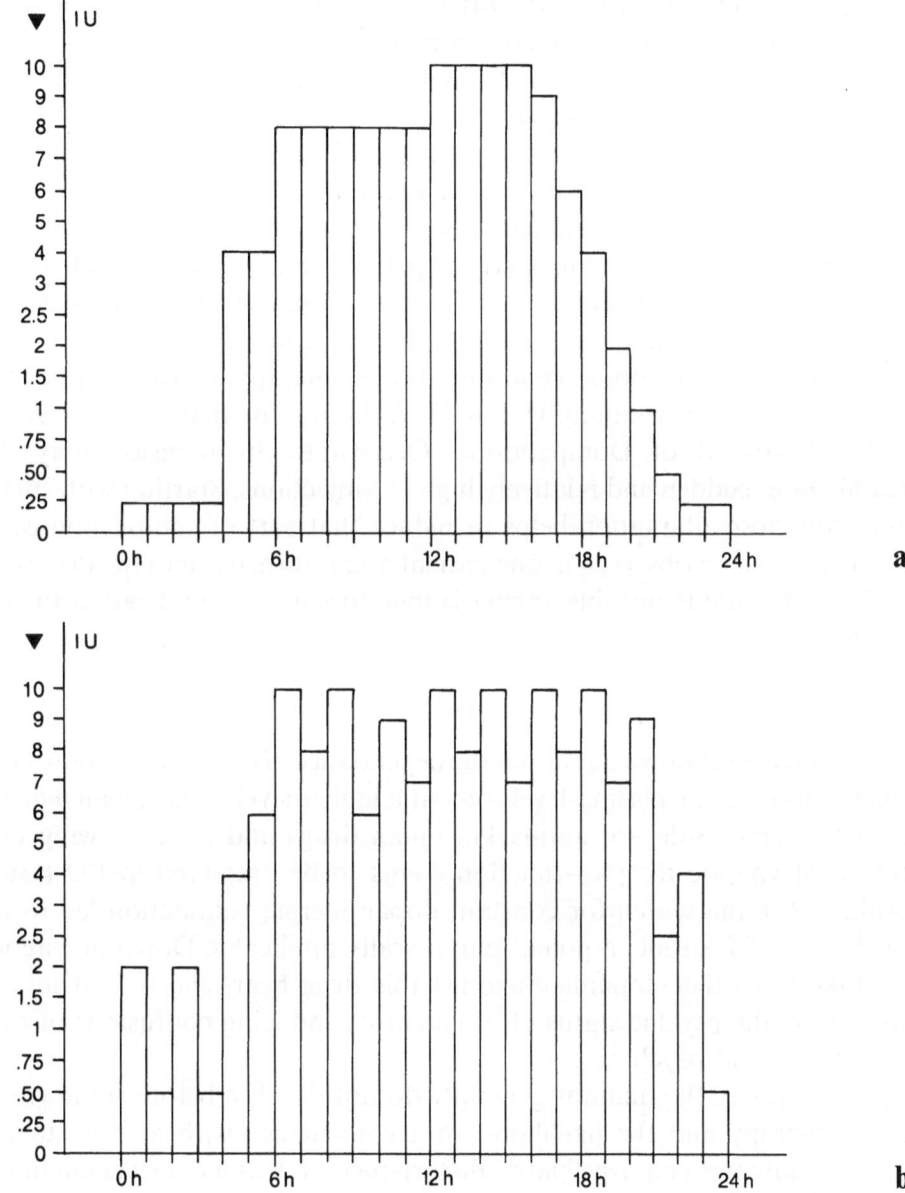

Fig. 6. More constant (**a**) or pulsatile (**b**) infusion-profile of s.c. Lisuride by the Disetronic-infusion pump

very anxious and with frequent episodes of overt psychosis, got psychotic on a routine hospital stay without any alteration of medication, day 1, 9:00 a.m. she got 15 mg Levomepromazine, at noon we stopped Lisuride-infusion (for her intolerable psychotic condition), at 7:00 p.m. on call she received 75 mg Melperone-HCl by the doctor, she slept well over night, day 2 at 5:00 a.m. she was taken to bath and reported to be well and at 8:30 a.m. she was found comatous with Cheyne-Stokes breathing, preserved reflexes and absent reaction on pain. We injected 0.2 mg Lisuride i.v. and 2–3 min afterwards the patient awoke, was fully alert and partly orientated, her breathing went back to normal, but she demonstrated retrograde amnesia for the past 12 hrs.

Confusion, illusions and hallucinations or agitated anxious status could be observed and responded well to dose-reduction in steps of about 10–20% of the daily dosis. In some cases 2–10 mg of Levomepromazine were given on acute conditions – it is worth-while mentioning that this very small dosis was effective in ameliorating the patients status.

S.c. nodules developed in nearly all patients and forced us only in one patient (H. M.) to stop infusion. On histologic examination there was unspecific inflammatory reaction and no proof of a specific allergic reaction.

Nausea could be observed at initiation of therapy in nearly all patients, but tolerance developed rapidly within 48–72 hours, for that time we recommend 3–4 – × 30 mg/d of Domperidone. Orthostatic hypotension may develop, mainly after sudden and relatively high i.v.-injections. Starting with small doses and slow dose adaptation helps to reduce that sort of complications. Angina pectoris could be observed in one patient and returned after repeated exposition to Lisuride. Stable/instable angina is therefore a contraindication for Lisuride therapy.

Discussion

With increasing knowledge about the dependance of even "wild" motor response fluctuations on the plasma-levels or—at a higher level, of speculated drug-levels at the receptor side—of antiparkinsonian drugs and after growing evidence, that postsynaptic receptor-function seems to be preserved in PD (Guttmann et al., 1985), the search for constant dopaminergic stimulation led to the idea of the use of Lisuride, a potent parenterally applicable Dopamine-agonist.

Like any other dopamine agonist this drug bears the risk of inducing or worsening the psychic status of a patient by inducing confusions or even hallucinations and psychoses.

As response fluctuations generally do not develop before some years of L-Dopa therapy and the likelihood of a coincidence with at this stage of the disease more frequent psychiatric disturbances, it was a clinical demand of high urgency to examine, whether a coincidence of both complications in longstanding PD still would be a suited entity for Lisuride therapy.

We therefore report on 5 Parkinsonian patients with previous psychiatric disturbances who suffered from severe motor-response fluctuations and who could not successfully be treated with oral therapy. The named psychiatric disturbances were repeated confusional episodes, illusions, hallucinations which in some cases neccessitated hospital care. All patients suffered from some sort of "on-off-fluctuations", which we define as sudden unpredictable variations of motility with no obvious correlation to the intake of antiparkinsonian-drugs (Barbeau, 1971; Castaigne et al., 1976; Marsden and Parkes, 1977). One patient was mostly akinetic, 2 patients had peak-dose hyperkinesias, a further patient some form of diphasic dyskinesia with predominant peak-dose akinesia. As patient or their relatives kept a motility diary for good, intermediate and bad motor response, we could calculate the reduction of off-periods under therapy by expressing the increase in on- and intermediate stages. For the small number of patients studied here, we generally renounced to make a further distinction between on- and intermediate phases.

Reasons for the parenteral application of Lisuride

There are various studies about the use of dopamine agonists in the reduction of on-off-fluctuations (Agid et al., 1979; Calne, 1982; Glantz et al., 1981; Hoehn, 1985; Lander et al., 1979; Marsden and Parkes, 1976 a; Parkes, 1979; Parkes et al., 1976; Rinne, 1983). In some patients these drugs would not be tolerated, or had to be reduced for psychiatric side effects (Barbeau, 1971; Marsden and Parkes, 1977; Moskowitz et al., 1978; Rinne, 1983; Teychenne et al., 1985).

After oral intake Bromocriptine is partially (ca. 34%) and Lisuride nearly fully (ca. 85%) absorbed, enteral absorption varies markedly and the first-pass-effect in the liver is rather high (85–95%, Burns and Calne, 1983). Thus plasma levels after oral intake of these drugs fluctuate markedly (Parkes, 1983) and we could achieve rather constant plasma levels after s.c. infusion of Lisuride which on top were reasonably higher than the correspondent plasma levels after oral intake (Figs. 7 and 8).

To our experience (and actually surprise) the constantly higher plasma levels of Lisuride were much better tolerated side-effect-wise (e.g. for nausea and psychotic reactions) than were much lower plasma levels with more pronounced fluctuations. One patient did not tolerate the nearly full reduction of Lisuride

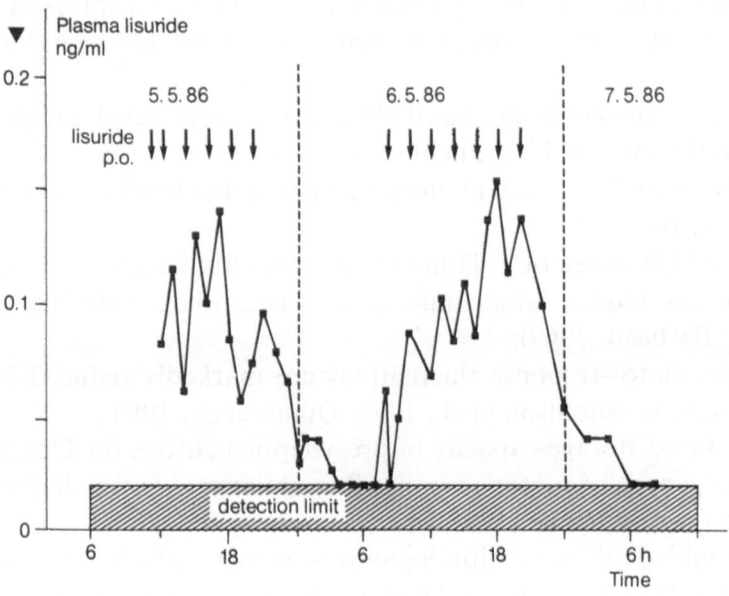

Fig. 7. Plasma-levels after oral intake of Lisuride (⬇: 0.2 mg tablet)

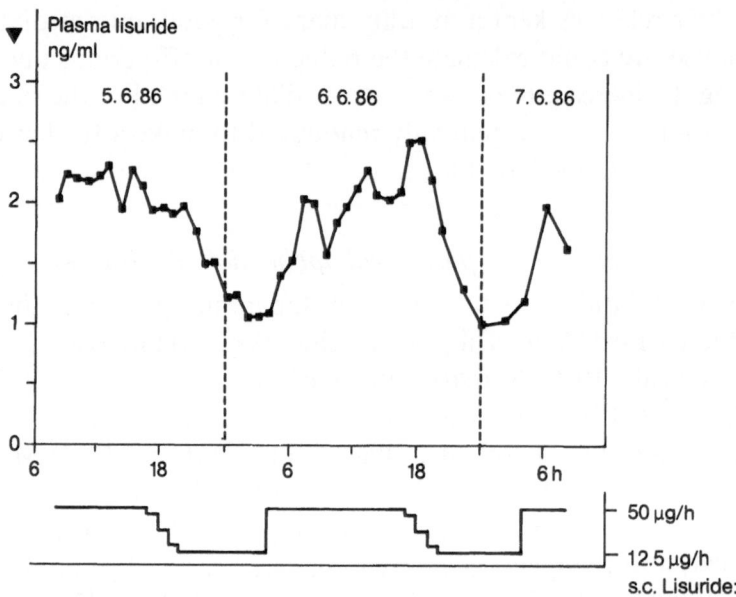

Fig. 8. Plasma-levels after s.c. infusion of Lisuride (12.5–50 µg/h)

at night-time and got better and less confused after increasing the dose of Lisuride at night-time to about 20–30% of the day-time dose.

The observed nausea after oral initiation of Lisuride generally vanishes within few days, because the D_2-receptors in the brainstem quickly develop tolerance (Dorow et al., 1983; Carruba et al., 1983). Nevertheless a sudden high plasma level of Lisuride again would cause nausea in some patients. One patient (H. M.) tolerated up to 2 mg Lisuride/d subcutaneously, but he would not tolerate any oral Lisuride, even not the 0.0125 mg tablets. In such a case we certainly have to discuss direct D_2-action on dopamine receptors in the gastro-enteral tract. This effect would be antagonized by Domperidone 3×20–3×30 mg.

For motor-neurons in the basal ganglia tolerance-development never has been shown (R. Horowski, oral comm.).

When we therefore look at motor-response fluctuations under L-Dopa-therapy, it seems
— that very high doses of L-Dopa (gram-amounts which have been used in the 1960's) may induce rather quickly motor-response disturbances (Cotzias et al., 1969; Barbeau, 1969);
— that these motor-response fluctuations are markedly reduced by constant L-Dopa-infusions (Shoulson et al., 1975; Quinn et al., 1984);
— that a reduced storage-capacity of presynaptic neurons for Dopamine plays a role in the genesis of motor-response fluctuations, as it was hypothesized by Parkes (1983) or by Rinne (1985 b);
— that a combined therapy with dopamine-agonists, which act at the postsynaptic receptor, helps to reduce L-Dopa intake (Calne, 1982; Calne and Stresse, 1985; Klawans et al., 1985; Rinne, 1985 a);

— that reduction of L-Dopa intake reduces metabolisation to 3-OMD, which for competition at the blood-brain-barrier with L-Dopa reduces the availability of L-Dopa in the brain (Muenter, 1973; Wade and Katzman, 1975; Gervas et al., 1983; Gerlach et al., 1986);

— that dopamine agonists may block the changes in dopamine receptor binding sites induced by levodopa (Rinne, 1985 a).

We have to admit that we know very little about the behavior of D_2 receptors in the limbic system. Whether any of the above described mechanisms plays a role in the action of these receptors still seems to be not clear. It is not even known, whether fluctuations in the psychic status of the patient are somehow an equivalent of motor response fluctuations. To examine this question, it may be helpful to keep Lisuride-plasma levels rather constant over a longer period, so that alterations in behavior actually might reflect changes in receptor sensitivity.

Motor response

Off-periods could be reduced by 54%, the hours of full or impaired mobility increased from 44.2% of the daytime to 74.4%. This reduction may not be as good as the results of Obeso et al. (1986), who reported a nearly full reduction of off-periods, but our figures were obtained in patients who tolerated only markedly smaller amounts of Lisuride (see below).

The amelioration of the clinical status is reflected as well by a decrease of the scores in the corresponding Parkinson's Disease Rating-Scales: in the Columbia University Rating-Scale the mean score decreased from 61.8 to 35.0 points, in the Kings's College P.D.R.S. the mean-scores decreased from 52.4 to 27.6 points for On-periods and from 88.6 to 64.8 points for Off-periods.

The last figures can be interpreted that even for the hours, when the patient judges himself "off", the quality and severity of these off-periods differs and that under the constant Lisuride infusion patients seem to be less impaired even when off.

An amelioration of the patients mood under Lisuride infusion corresponds to a decrease in the ZUNG's depression score (range 20–80 points, reduction from 50.6 to 38.6 points).

There was not only an increase of On-periods; especially those patients with dyskinesias or peak-dose hyperkinesias had great benefit from the constant dopaminergic infusion and we could achieve a total abolition of these involuntary movements, in no case we could observe choreatic movements. Our extensive reduction of L-Dopa-intake certainly helped very much in avoiding these unwanted motor-effects.

Total daily dose of Lisuride

We gradually increased Lisuride up to 2 mg/d, but within few weeks the maximum dosage our patients would tolerate was only 0.94 ± 0.65 Lisuride/d. The Spanish group used 2–4 mg of Lisuride over a rather long period with less psychiatric disturbances (Obeso et al., 1986). Possible reasons for that difference could be the sort of patients studied: with a mean duration of the disease of 13.4 years, of established anti-Parkinson therapy for 10.8 years and the described psychiatric disturbances our patients would not tolerate a much higher intake of a dopamine-agonist.

Concomittant reduction of L-Dopa

When establishing the Lisuride-pump, we immediately reduced L-Dopa intake. With 0.5 mg Lisuride/d we could reduce L-Dopa intake by 20–30%, with maximum Lisuride (1.5 mg/d) we reduced L-Dopa by 40–65%. In the patients described here, L-Dopa was reduced from 1025 ± 495 to 362 ± 219 mg whilst Lisuride was given with a mean infusion rate of 0.94 ± 0.65 mg/d.

There is certainly a vast tolerance for personal preferences of how to reduce L-Dopa and how to get a well balanced relation between L-Dopa and Lisuride. But in nearly any case we would recommend a significant reduction of L-Dopa when establishing the Lisuride-pump, otherwise the probability of psychiatric disturbances like confusional states or even psychotic events is far too high. As psychiatric disturbances may also develop in patients who beforehand never had any symptoms of that sort, and for that these complications may show up rather unexpected and without any sign of warning we think that a reduction of L-Dopa is obligatory. As receptor-sensitivity seems on top to alter to "hypersensitivity" within the first 8–12 weeks of treatment (see below), we think it is favourable to accept some moderate akinesia for a few weeks and if necessary then to increase either L-Dopa or the dopamine agonist than to accept the high and possibly vital risk of overt psychosis with the only advantage of immediate improvement of mobility.

The advantages of a reduction of L-Dopa-intake seem to be evident, one might name
— reduction of peak-dose hyper- or dyskinesias;
— decrease of L-Dopa-metabolites like 3-OMD;
— less systemic side-effects (e.g. gastro-intestinal);
— reduced probability of generating late complications of high-dose L-Dopa therapy like motor-response fluctuations (Cotzias et al., 1970; Barbeau, 1976, 1980; Rinne, 1985 b; Claveria et al., 1973);
— remaining allowance for further increase of L-Dopa in the following years;
— smaller probability of psychiatric disturbances.

For such cases where Lisuride should be applied without any L-Dopa intake it seems that Lisuride alone is not very effective in about 20% of Parkinson-patients (Rinne, oral communication).

Diurnal alteration of the Lisuride-infusion-profile

We tried to establish a rather high plasma-level in the morning ("loading dose"), lowering the infusion rate at midday and increasing the rate again in the afternoon, when motility generally gets worse and the demand for dopaminergic stimulation increases. We then gradually decreased the infusion-rate towards the evening so that the plasma level of Lisuride steadily would fall down at night time. A full reduction of Lisuride at night seems to be less well tolerated than a reduced infusion rate of 20–30% of the day-time infusion rate (see above).

A possible explanation for this could be the partial serotonergic agonistic properties of Lisuride (Pieri et al., 1983; Carvey and Klawans, 1983). Later on we altered the above described "M-shape" of the infusion profile by letting the infusion rate fluctuate by about 20–30% around the named baseline (Fig. 6 a, b). To our own astonishment this alteration of the infusion profile was followed by a marked improvement in the mobility without unwanted additional psychotic side-effects. We are presently examining in a double-blind study, whether this hourly changing profile is actually better tolerated than a more steady infusion profile. A first open trial confirmed a marked deterioration in one patient, whom we changed from the hourly changing profile to the more uniform infusion-profile. We are examining as well, whether more pulsatile infusion rates (changing every 10 min) possibly are even better tolerated or whether our first impression was either casual or even has to be rejected.

Changes of receptor-sensitivity

For repeatedly developing psychiatric disturbances after continuation of a primarily well tolerated dose of Lisuride, we reduced Lisuride in severel steps within the first 8–12 weeks to a certain level, which then could generally be maintained. Later episodes of "intolerance" might relate to the so called long-term fluctuations of sensitivity, described as day by day or week by week fluctuations (de Jong et al., 1962) or long-term clinical fluctuations (Fahn, 1974). These fluctuations made (even after some months) temporary dose reductions of Lisuride necessary, but later-on the dose could be increased again to the previous level.

Whether the observed necessity for a reduction of Lisuride within the first three months actually reflects a developing hypersensitivity of postsynaptic neurons is not proven (for comparison, see Teychenne et al., 1985). At least it seems not clear, whether the observed "hypersensivitiy" is a direct effect of Lisuride stimulation and/or whether the concomittant dose-reduction of L-Dopa and thereby a certain decrease of 3-OMD and possibly of other metabolites plays a role (this would be comparable to the observed amelioration of the clinical status after so-called "drug-holidays").

In some further patients, in whom we used Lisuride infusions it seemed, that keeping a constant (and from the beginning smaller) dose of Lisuride, that would not have to be reduced within the first weeks motor-response would ameliorate after some weeks. This would coincide with the observation of Obeso et al. (1986), that chorea was not observed within the first 3–4 weeks but appeared in all three patients described there during the second or third month. The authors discussed just so the emergence of dopaminergic sensitivity.

Local tolerance

As s.c. nodules emerged in some patients about 12–36 hrs after every change of the injection site, we used different needles of metal or plastic material to examine their influence on nodule-formation.

In 2 patients, who developed nodules of over 0.5 cm diameter after metal needles, the nodule formation was reduced nearly to zero with the change to plastic needles like Abbocath® or Microcath®. But even amongst different metal-needles there seemed to be some difference in the size of s.c. nodules. So it might be possible, that the observed toxic-allergic reaction may be caused by the metal of the alloy dissolved under the influence of the slightly acid Lisuride-solution.

Further indications

We just want to mention here some further indications of i.v./s.c. Lisuride:

Neuroleptic malignant syndrome (NMS)

We have treated 4 patients with the clinical features of that syndrome – in one patient therapy was necessary for more than 4 weeks and reduction of Lisuride after 2 weeks was followed again by the development of akinesia. Lisuride was given in doses of up to 6 mg/d (publication in preparation).

Akinetic crisis

Two patients with PD had been treated in another hospital with neuroleptics and were akinetic, they responded well on 0.5–1 mg Lisuride for 8–12 days, their akinesia resolved and they could be sufficiently treated by further oral medication.

Tardive dyskinesias

Two patients with neuroleptic-induced tardive dyskinesia were treated successfully with the Lisuride-pump, one patient (65 years, tardive dyskinesia for 8 years) had a decrease of her symptoms by estimated 60–70% under 0,45 mg Lisuride/day, no other oral treatment was that effective, and even oral Lisuride we tried in her was not half as effective as s.c. infusion, therefore continous infusion is now maintained for more than 15 months.

Torticollis spasmodicus

Four patients with long standing torticollis and with no response to any of the usual oral medication responded well to the continous s.c. Lisuride infusion. In three patients there was already severe dystonia of the trunc with axial skeleton deformity, one patient was immobile and needed to be attended. In these patients the dystonia markedly was diminished and the patients could be mobilised well and remain under that therapy.

Conclusion

We conclude that the constant s.c. Lisuride-infusion may be very effective in reducing on-off motor response fluctuations in Parkinsonian patients. As Lisuride is a potent dopamine agonist, this drug can dose-dependently induce psychosis or confusional states in these patients, who anyway are at high risk to develop these complications or dopaminergic therapy. By reducing the intake of L-Dopa (+ DCI) as well as of other anti-parkinsonian drugs, best by avoiding any additional intake of anticholinergics which seem to bear a rather high risk of inducing these unwanted side effects and by a very subtle and time-extensive adjustment of the infusion profile of Lisuride to the individual need of the patient unwanted psychiatric side effects may be minimized. Within the first 3 months of therapy dose adjustment of Lisuride has carefully to be undertaken, as within this period there seems to be a change in receptor-sensitivity to hypersensitivity which increases the risk for developing psychiatric side-effects.

In patients who besides these precautions still developed confusions or hallucinations, we used the neuroleptic Levomepromazine in dosis of 5 mg–25 mg/d, but as this drug worsens the motor response in these patients, we meanwhile treat these patients with 25 mg–50 mg of the neuroleptic Clozapine, which nearly does not alter motor performance in these patients. For reported agranulocytosis under Clozapine (Amsler et al., 1977) regular blood counts have to be done even with small doses of this agent.

A sudden stop of constant Lisuride infusion bears the risk of a dopamine − deficiency syndrome similar to the neuroleptic malignant syndrome. Patients and their relatives should be well informed for the possible occurrence and treatment of these complications.

The emergence of s.c. nodules is greatly reduced by the application of plastic needles (Abbocath®, Microcath®) instead of metal needles.

Lisuride is a potent dopaminergic agent and seems to be a good candidate for the parenteral treatment of dopamine-deficiency syndromes as acute akinetic crisis, the neuroleptic malignant syndrome and related conditions as well as for dyskinesias, whether neuroleptic-induced (tardive dyskinesias) or of unknown etiology like torticollis spasmodicus.

References

Agid Y, Bonnet AM, Pollak P, Signoret JL, Lhermitte F (1979) Bromocriptine associated with a peripheral dopamine blocking agent in treatment of Parkinson's disease. Lancet 1: 570–572

Amsler HA, Teerenhovi L, Barth E, Harjula K, Vuopio P (1977) Agranulocytosis in patients treated with clozapine. A study of the Finnish epidemic. Acta Psychiatr Scand 56: 241–248

Barbeau A (1969) L-Dopa therapy in Parkinson's disease. A critical review of 9 years experience. Can Med Assoc J 101: 791–800

Barbeau A (1971) Long-term side effects of levodopa. Lancet 1: 395

Barbeau A (1976) Pathophysiology of oscillations in performance after long-term therapy with L-Dopa. In: Birkmayer W, Hornykiewicz O (eds) Advances in Parkinsonism. Editiones Roche, Basel, pp 424–434

Barbeau A (1980) High level levodopa therapy in severely akinetic parkinsonian patients: twelve years later. In: Rinne UK, Klingler M, Stamm G (eds) Parkinson's disease. Current progress, problems, and management. Elsevier, Amsterdam, pp 229–239

Birkmayer W (1978) Long term treatment with L-Deprenyl. J Neural Transm 43: 239–244

Birkmayer W, Knoll J, Riederer P, Youdim MBH (1983) (—)Deprenyl leads to prolongation of L-Dopa efficacy in Parkinson's disease. Mod Probl Pharmacopsychiatry 19: 170–176

Birkmayer W, Riederer P (1975) Responsibility of extrastriatal areas for the appearance of psychotic symptoms. Clinical and biochemical human post-mortem findings. J Neural Transm 37: 175–182

Birkmayer W, Riederer P, Abrozi L, Youdim MBH (1977) Implications of combined treatment with Madopar® and L-deprenyl in Parkinson's disease. Lancet 1: 439–443

Bittkau S, Przuntek H (1986) Lisuride infusion pump for PD. Lancet 2: 349

Burns RS, Calne DB (1983) Disposition of dopaminergic ergot compounds following oral administration. In: Calne DB, Horowski R, McDonald RJ, Wuttke W (eds) Lisuride and other dopamine agonists. Raven Press, New York, pp 153–160

Calne DB (1982) Dopamine receptor agonists in the treatment of basal ganglia disorders. Semin Neurol 2: 359–364

Calne DB, Claveria LE, Allen JG (1974) Plasma levodopa and the on/off effect. Can-Am Conf on PD. Raven Press, New York, Adv Neurol 5: 341–344

Calne DB, Stresse J (1985) Approaches to the use of Bromocriptine in Parkinson's disease. In: Fahn S, Marsden CD, Jenner P, Teychenne PF (eds) Approaches to the use of Bromocriptine in Parkinson's disease. Raven Press, New York, pp 255–258

Carruba MO, Mantegazza P (1983) Behavioral pharmacology of Ergot derivatives. In: Calne DB, Horowski R, McDonald RJ, Wuttke W (eds) Lisuride and other dopamine agonists. Raven Press, New York, pp 65–77

Carvey PM, Klawans HK (1983) Effect of chronic Lisuride treatment on stereotypical and myoclonic jumping behavior in Guinea pigs. In: Calne DB, Horowski R, McDonald RJ, Wuttke W (eds) Lisuride and other dopamine agonists. Raven Press, New York, pp 97–107

Castaigne P, Rondot P, Ribadeau-Dumas JL, Cardon P (1976) Long-term treatment of Parkinson's disease with L-Dopa alone and in combination with a decarboxylase inhibitor. In: Birkmayer W, Hornykiewicz O (eds) Advances in Parkinsonism. Editiones Roche, Basel, pp 462–468

Claveria LE, Calne DB, Allen JG (1973) On-off phenomenon related to high plasma levodopa. Br Med J 2: 641–643

Cotzias GC, Papavasiliou, Gellen (1969) Modification of Parkinsonism. N Engl J Med 280: 337–345

Cotzias GC, Papavasiliou PS, Steck A, Duby S (1970) Parkinsonism and levodopa. Clin Pharmacol Ther 12: 319–322

Critchley P, Grandas PF, Quinn N, Coleman R, Parkes D, Marsden CD (1986) Psychosis and the Lisuride pump. Lancet 2: 349

Curzon G, Friedel J, Grier L, Marsden CD, Parkes JD, Shipley M, Zilkha KJ (1973) Sustained-release levodopa in parkinsonism. Lancet 1: 781

De Jong J, David J, Burns BD (1962) An investigation of Parkinson's disease. Neurology 12: 402–409

Dorow R, Breitkopf M, Gräf KJ, Horowski R (1983) Neuroendocrine effects of Lisuride and its 9,10-dihydrogenated analog in healthy volunteers. In: Calne DB, Horowski R, McDonald RJ, Wuttke W (eds) Lisuride and other dopamine agonists. Raven Press, New York, pp 161–174

Dubois R (1986) On-off phenomenon: description and incidence. In: Marsden CD, Rinne UK, Koella W, Dubois R (eds) International workshop on the on-off phenomenon in Parkinson's disease. Eur Neurol [Suppl] 27: 3–8

Eckstein B, Shaw K, Stern G (1973) Sustained-release levodopa in parkinsonism. Lancet 1: 431–432

Eriksson T, Magnusson T, Carlsson A, Linde A, Granérus AK (1984) "On-Off" phenomenon in Parkinson's disease: correlation to the concentration of Dopa in plasma. J Neural Transm 59: 229–240

Fahn S (1974) On/off phenomenon with levodopa therapy in Parkinsonism. Clinical and pharmacological correlations and the effect of intramuscular pyridoxine. Neurology 24: 431–441

Gerlach M, Gebhardt B, Kuhn W, Przuntek H (1986) Die Abhängigkeit der Resorption des L-Dopa von Galenik und veränderter Magensaftsekretion des Parkinson-Patienten. In: Fischer PA (ed) Spätsyndrome der Parkinson'schen Krankheit. Editiones Roche, Basel, pp 271–279

Gervass JJ, Muradas V, Bazan E (1983) Effects of 3-OMD on monoamine metabolism in rat brain. Neurology 33: 278–282

Glantz R, Goetz CG, Nausieda PA (1981) The effect of bromocriptine (BCT) on the on-off phenomenon. J Neural Transm 52: 41–47

Guttmann M, Seeman P, Reynolds GP, Riederer P, Jellinger K, Tourtellotte WW (1986) Dopamine D_2 receptor density remains constant in treated Parkinson's disease. Ann Neurol 19: 487–492

Hardie RJ, Lees Aj, Stern GM (1984) On-off fluctuations in Parkinson's disease. Brain 107: 487–506

Hoehn MM (1985) Using bromocriptine to manage the inadequacies of levodopa therapy. In: Fahn S, Marsden CD, Jenner P, Teychenne P (eds) Approaches to the use of bromocriptine in Parkinson's disease. Raven Press, New York, pp 46–58

Klawans HL, Goetz CG, Tanner CM, Glantz R (1985) Studies with Bromocriptine. In: Fahn S, Marsden CD, Jenner P, Teychenne PF (eds) Approaches to the use of Bromocriptine in Parkinson's disease. Raven Press, New York, pp 37–45

Lander CM, Lees A, Stern G (1979) Oscillations in performance in levodopa-treated Parkinsonians: treatment with bromocriptine and L-deprenyl. Clin Exp Neurol 16: 197–203

Marsden CD, Parkes JD (1976 a) Bromocriptine in parkinsonism. Lancet ii: 419–420

Marsden CD, Parkes JD (1976 b) "On-off" effects in patients with Parkinson's disease on chronic levodopa therapy. Lancet i: 292–296

Marsden CD, Parkes JD (1977) Success and problems of long-term levodopa therapy in Parkinson's disease. Lancet ii: 345–349

Moskowitz C, More H, Klawans H (1978) Levodopa induced psychosis: a kindling phenomenon. Am J Psychiatry 135: 669–675

Muenter MD, Dinapoli RP, Sharpless NS, Tyce GM (1973) 3-O-Methyldopa, L-Dopa, and trihexyphenidyl in the treatment of Parkinson's disease. Mayo Clin Proc 48: 173–183

Nutt JG, Woodward WR (1986) Levodopa pharmacokinetics and pharmacodynamics in fluctuating parkinsonian patients. Neurology 36: 739–744

Nutt JG, Woodward WR, Hammsterstad JP, Carter JH, Anderson JL (1984) The "on-off" phenomenon in Parkinson's disease. Relation to Levodopa absorption and transport. N Engl J Med 310: 483–488

Obeso JA, Luquin MR, Martinez-Lage JM (1986) Lisuride infusion pump: a device for the treatment of motor fluctuations in Parkinson's disease. Lancet i: 467–470

Parkes JD (1979) Bromocriptine in the treatment of parkinsonism. Drugs 17: 365–382

Parkes JD (1983) Variability in Parksinson's disease; clinical aspects, causes and treatment. Acta Neurol Scand 95: 27–35

Parkes JD, Marsden CD, Donaldson I (1976) Bromocriptine treatment in Parkinson's disease. J Neurol Neurosurg Psychiatry 39: 184–193

Pieri L, Keller HH, Laurent JP, Burkard WP, Pieri M, Bonetti EP, Da Prada M (1983) Behavioral, neurochemical, and electrophysiological effects of Lisuride and LSD in animals. In: Calne DB, Horowski R, McDonald RJ, Wuttke W (eds) Lisuride and other dopamine agonists. Raven Press, New York, pp 89–96

Poewe WH, Lees AJ, Stern GM (1987) Clinical and pharmakokinetic observations with Madopar HBS® in hospitalized patients with Parkinson's disease and motor fluctuations. Eur Neurol 27 [Suppl 1]: 93–97

Quinn N, Parkes JD, Marsden CD (1984) Control of on/off phenomenon by continuous intravenous infusion of Levodopa. Neurology 34: 1131–1136

Rinne UK (1985 a) Early Combination of Bromocriptine and Levodopa in the treatment of Parkinson's disease. In: Fahn S, Marsden CD, Jenner P, Teychenne PF (eds) Approaches to the use of Bromocriptine in Parkinson's disease. Raven Press, New York, pp 7–14

Rinne UK (1985 b) Madopar HBS in the long-term treatment of Parkinsonian patients with fluctuations in disability. Abstract. Eur Neurol 27 [Suppl 1]: 120–124

Rinne UK (1983 a) New ergot derivatives in the treatment of Parkinson's disease. In: Calne DB, Horowski R, McDonald RJ, Wuttke W (eds) Lisuride and other dopamine agonists. Raven Press, New York, pp 431–442

Rinne UK (1983 b) Problems associated with long-term levodopa treatment of Parkinson's disease. Acta Neurol Scand 95: 19–26

Rinne UK, Siirtola T, Sonninen V (1978) L-Deprenyl treatment of on-off-phenomena in Parkinson's disease. J Neural Transm 43: 253–262

Rivera-Calmlim L, Tandon D, Anderson F, Joynt R (1977) The clinical picture and plasma levodopa metabolite profile of Parkinsonian nonresponders. Arch Neurol 34: 228–232

Shoulson I, Glaubiger GA, Chase TN (1975) On-off response: clinical and biochemical correlations during oral and intravenous levodopa administration in Parkinsonian patients. Neurology 25: 1144–1148

Sweet RD, McDowell FH (1974) Plasma dopa concentrations and the "on-off" effect after chronic treatment of PD. Neurology 24: 953–956

Teychenne PF, Bergstrud D, Racy A (1985) Subcategories of Parkinson patients treated with Bromocriptin. In: Fahn S, Marsden CD, Jenner P, Teychenne P (eds) Approaches to the use of Bromocriptin in Parkinson's disease. Raven Press, New York, pp 59–72

Todes CJ (1986) At the receiving end of the Lisuride-pump. Lancet ii: 36–37

Tolosa ES, Martin WE, Cohen HP, Jacobson RL (1975) Patterns of clinical response and Plasma Dopa-levels in Parkinson's disease. Neurology 25: 177–183

Wade LA, Katzman R (1975) 3-O-methyldopa uptake and inhibition of L-dopa at the blood-brain barrier. Life Sci 17: 131–136

Wajsbort J, Kartmazov K, Oppenheim B, Barkey R, Youdim MBH (1982) The clinical and biochemical investigation of L-Deprenyl in Parkinson's disease with special reference to the "on-off" effect. J Neural Transm 55: 201–215

Authors' address: Dr. S. Bittkau, Neurologische Universitätsklinik, Josef Schneider-Strasse 11, D-8700 Würzburg, Federal Republic of Germany.

J Neural Transm (1988) [Suppl] 27: 55–60

Continuous subcutaneous lisuride infusions in Parkinson's disease

P. H. S. Critchley, F. Grandas Perez, N. P. Quinn, J. D. Parkes,
and C. D. Marsden

University Department of Neurology and the Parkinson's Disease Society Research Centre,
King's College School of Medicine and Dentistry and Institute of Psychiatry,
London, U.K.

Summary. Thirteen patients with idiopathic Parkinson's disease and "on-off" fluctuations on oral levodopa plus dopa decarboxylase inhibitor (DDI) were treated with continuous (24 hour) subuctaneous lisuride infusions together with a reduced dose of levodopa (plus DDI). An improvement in motor performance was seen in 10 patients, with a mean increase in percentage of waking time spent "on" of 32 per cent (range 13–59 percent). However, adverse effects were common, especially psychiatric effects, leading to treatment withdrawal in 11 of 13 subjects after a mean of 40 days' treatment.

Continuous lisuride infusion together with a small dose of levodopa (plus DDI) are effective treatment for "on-off" fluctuations in Parkinson's disease, but the frequency of adverse effects limits the number of patients who can be treated successfully with this technique.

Introduction

Fluctuations in motor performance develop in up to 50 percent of patients with Parkinson's disease treated with levodopa for between two and five years (Marsden and Parkes, 1976). In the majority of such patients, increasing the frequency and decreasing the size of individual doses of levodopa improves the motor performance, but often for a limited period only. Thus there is a group of people in whom alterations in timing and/or size of individual doses of levodopa, or the addition of other antiparkinsonian drugs, give little additional benefit. Such patients with "resistant" fluctuations respond well to a constant intravenous infusion of levodopa which, when given in a large enough dose, can abolish "off" periods (Nutt et al., 1984; Quinn et al., 1984; Hardie et al., 1984). However, levodopa needs to be given in large volumes (two-three litres/day) because of the low pH of the solution, and thus is impractical for use in the patient's home. A more practical approach to this problem would be a drug which was highly soluble and could also be given subcutaneously, using portable mini-pumps developed for the administration of insulin. Lisuride, a directly-acting D 2 dopamine agonist, fulfils these requirements. Obeso et al. (1983) were the first to show the sustained antiparkinsonian effect of lisuride when given by intravenous infusion (diluted in 500 ml of saline) over 8 h. The same group went on to show a sustained antiparkinsonian effect of lisuride over many months, given by continuous (24 h) subcutaneous infusion together with small doses of levodopa (plus DDI) (Obeso et al., 1986). In this latter study, lisuride was given in very small volumes of saline (at a concentration of up to 1 mg/ml, up to 2 ml/day), allowing patients to remain in their own homes.

Others have reported similar beneficial results, but adverse effects have been a problem (Ruggieri et al., 1986; Bittkau and Przuntek, 1986). At the Fourth European Workshop on Clinical Neuropharmacology in 1986, results were reported in 60 patients treated with lisuride infusions. Twenty-six developed psychiatric effects, requiring withdrawal of lisuride treatment in approximately half of them.

We describe here our experience with lisuride infusions in 13 patients with Parkinson's disease.

Material and methods

Thirteen patients (ten men and three women) with idiopathic Parkinson's disease and "on-off" fluctuations in motor performance on levodopa (plus DDI) were treated with subcutaneous lisuride given by continuous 24-hour infusion. All patients had experienced choreiform dyskinesias associated with "on" periods whilst on levodopa, but one was experiencing "on" periods without involuntary movements immediately prior to the study. Four patients had in addition experienced severe painful early morning dystonic spasms before being treated with lithium (Quinn et al., 1986). Three other patients had a times experienced mild dystonic spasms. Their mean age was 55 years (range 32–70); mean duration of symptoms 14 years (range 7–20). Thus they constituted a group of subjects with relatively early disease onset, mean 41 years (range 25–63). Such patients are known to be particularly at risk of the precocious development of severe involuntary movements and fluctuations in motor performance (Quinn et al., in preparation). Severity of parkinsonism was assessed by the King's College Hospital Parkinson's Disease Clinic rating scale (Quinn et al., 1982); mean disability score "on" was 21 (range 3–45), and "off" 58 (range 45–76). Mean daily levodopa dosage (plus DDI) was 1394 mg/day (range 600–3150), given in a mean of 9.6 doses (range 5–20). Eight patients were on additional antiparkinsonian medication. This included selegiline, four; benzhexol, two; pergolide, one; and bromocriptine, one. In addition, four were taking lithium carbonate as a treatment for "off" period dystonic spasms.

All subjects were given domperidone 20 mg tds po prophylactically against nausea and vomiting. Lisuride, given by continuous (24 h) subcutaneous (sc) infusion, was added to the other antiparkinsonian drugs. The initial dose of lisuride given was 16 micrograms/h. This was gradually increased (decreased in one) as the levodopa was decreased and other antiparkinsonian drugs withdrawn, until optimal beneficial effects were obtained. The mean maintenance dose of lisuride was 55 micrograms/h (range 8–117), i.e. 1.3 mg/day. The levodopa dose was decreased from a mean of 1394 mg/day to 785 mg/day, i.e. 56 percent of the former dose, given in a mean of nine doses per day (range 5–20).

"On-off" charts were kept by each patient for three to seven days before the lisuride was started, and throughout the period of lisuride infusion whilst in hospital. In addition, severity of dyskinesias was self-rated as mild, moderate or severe. Pulse, and blood pressure lying and standing, were measured four times per day.

The means of the numbers of "off" periods per day before and during lisuride treatment were compared using a paired t test.

Results

Motor

At the time of optimum benefit from lisuride, motor performance, as determined by an increase in percentage waking time spent on, was improved in 10 out of 13 patients. In these responders, the mean increase in percentage of waking time "on" was 32 percent (range 13–59 percent). This improvement was maintained for periods of up to 128 days, until adverse effects developed and treatment had to be withdrawn. The mean duration of treatment was 40 days. In the three subjects in whom no improvement in motor performance was seen, adverse effects developed (after 36 hours, 10 and 11 days respectively) before a steady state of lisuride infusion was reached. The number of "off" periods per day was decreased from a mean of 4.5 + 0.5 (mean ± SEM) to a mean of 3.9 ± 0.3 (t = 0.97, df 12, p > 0.05) a non-significant change. "Off" periods were less severe; one patient previously unable to walk in "off" periods was able to do so on lisuride treatment.

Adverse effects

Adverse effects were common, and were responsible for withdrawal of treatment in $^{11}/_{13}$ patients. In eight this was because of psychiatric effects alone; in two because of psychiatric effects and an increase in involuntary movements; and in one because of psychiatric effects and an increase in postural hypotension.

Psychiatric effects

Psychiatric effects occurred in 11 patients. They were of gradual onset, evolving through a period of agitation and increased sensory awareness to a paranoid state, later accompanied by confusion. There was a marked diurnal variation in the severity of psychiatric symptoms, with a relatively normal mental state in the mornings, followed by a deterioration in the evening and night. Paranoid delusions (eight) included morbid jealousy and ideas of persecution. An organic confusional state was seen in seven patients. Visual hallucinations (five) included soldiers on sentry duty and shadowy burglars; auditory hallucinations (four) included conversations about health conducted in the third person, and sounds made by people moving about upstairs. Hypersexuality was seen in one.

In at least one patient the paranoid delusional state seen before confusion developed was considered by the attending psychiatrist (Dr. B. Toone) to be indistinguishable from an acute paranoid schizophrenic psychosis.

In an attempt to overcome these psychiatric effects in four patients, the lisuride dose was decreased by 25 percent. However this did not lead to improvement in the mental state, and lisuride had to be withdrawn. In four other patients, stopping the lisuride until the psychiatric effects resolved (after 4–10 days), and then reintroducing lisuride at half the former dose, resulted in a recurrence of psychiatric effects over a shorter period than previously seen.

Involuntary movements

Involuntary movements, which were mainly generalized and choreic, and seen throughout "on" periods, were increased in seven patients. They were of the same distribution and nature as previously seen in the same patient on their former antiparkinsonian therapy. Adjustments in the dose of lisuride and levodopa reduced the involuntary movements to acceptable, i.e. former, levels in five of these patients, but in the other two increased involuntary movements contributed to withdrawal of lisuride treatment.

Nausea

Nausea occurred in three subjects; vomiting in one. These effects were all transient.

Subcutaneous nodules

Subcutaneous nodules developed at the injection sites in all patients treated for more than 12 days. Injection sites were initially changed every four days, but nodules were not seen until the third or fourth change. Subsequently injection sites were changed every two to three days, because of nodules developing earlier, and also clinical benefit being less well sustained on the fourth day after a needle change. The nodules were small (diameter approximately five mm), round and erythematous. They were occasionally itchy, but not painful. The nodules resolved gradually over 7–14 days following withdrawal of the needle from that site. In one patient at one site on one occasion only, a subcutaneous abscess developed. This resolved rapidly after the pus was expressed. This patient is currently tolerating subcutaneous needles well.

Hypotension

Clinically significant postural hypotension occurred in only one patient. This patient had an autonomic neuropathy, which was first manifest 12 years after Parkinson's disease was diagnosed. Prior to the study, her hypotension on levodopa had responded to fludrocortisone 0.1 mg/day. During the course of lisuride infusion, postural hypotension was not initially troublesome, but became so after 50 days' treatment. An increase in fludrocortisone dose at this time did not improve the symptoms and, after a further 10 days of lisuride treatment, she collapsed briefly with profound hypotension, developing at the same time paranoid delusions with confusion. Lisuride treatment was withdrawn at this time.

Discussion

The benefits of D-L-dopa treatment in Parkinson's disease were first convincingly demonstrated by Cotzias et al. (1967, 1969). Nineteen years on, L-dopa is still the drug of first choice in the treatment of Parkinson's disease. However, the problem of "on-off" fluctuations (Marsden and Parkes, 1976) has led to the search for approaches to treatment other than oral levodopa. Data obtained from levodopa infusion work show that a constant plasma level of levodopa, and therefore presumably a more constant striatal delivery of dopamine than can be obtained with oral treatment, can produce a smooth motor response in patients with "on-off" fluctuations on oral levodopa treatment (Quinn et al., 1984; Nutt et al., 1984; Hardie et al., 1984). Likewise, lisuride given by continuous subcutaneous infusion can produce a relatively constant plasma level of lisuride (Obeso, personal communication). In the current study, however, optimal beneficial results from lisuride infusion also required additional small doses of oral levodopa (plus DDI). The reason why the combination of lisuride and levodopa is more effective than either drug alone is not clear. Perhaps the explanation is that a favourable response to combined lisuride/levodopa therapy depends in part on pharmacokinetic factors, as well as the somewhat different adverse effect of the two drugs, levodopa being more likely to cause dose-limiting involuntary movements, while lisuride is more likely to cause frequent psychiatric problems.

In the short term, the benefits from the lisuride/levodopa combination were greater than those obtained from previously optimized oral treatment in the same patient. However, sustained abolition of "off" periods was not obtained, and a compromise was necessary between adverse and beneficial effects. Only two patients received lisuride treatment for longer than three months. One gradually lost the beneficial effects whilst becoming psychotically depressed with associated morbid jealousy, the other lost beneficial effects following a 20 percent dosage decrease which became necessary when he developed visual illusions.

Psychiatric effects occurred in 11 subjects, 10 of whom had previously experienced psychiatric effects on other antiparkinsonian drugs before the lisuride study. In addition, one of the 10 had previously had a depressive illness, and one other showed evidence of cognitive impairment. Cognitive impairment, and a previous psychiatric illness, are recognized as risk factors for the subsequent development of psychiatric effects on antiparkinsonian drugs (Moskovitz et al., 1978). In the context of the current study, a previous episode of psychiatric effects on other antiparkinsonian drugs is also a risk factor for psychiatric effects occurring with the lisuride/levodopa combination. Thus, patients with a previous psychiatric episode on antiparkinsonian drugs will in future not be given long-term parenteral lisuride treatment.

All the psychiatric effects seen in this study have been described as occurring on levodopa alone (Goodwin, 1971). This suggests that the mechanism for these psychiatric effects is dopaminergic, and that the serotoninergic effects of lisuride are thus not important in this regard. Furthermore, 5 HT activity, as determined in animal studies with ergot derivatives does not accurately predict the hallucinogenic potential of these drugs in normal human volunteers (Fanchamps, 1978).

Thus the chronic subcutaneous infusion of lisuride reduces disability from motor response fluctuations in patients with Parkinson's disease. However, in our experience, only a minority of patients can tolerate such treatment. The general principle of continuous dopaminergic stimulation is however well established, and needs to be extended to include other drugs with a lower potential for causing psychiatric effects, and also other routes of administration, including oral and transdermal routes.

References

Bittkau S, Przuntek H (1986) Lisuride infusion pump for Parkinson's disease. Lancet 2: 349

Fanchamps A (1978) Some compounds with hallucinogenic activity. In: Berde B, Schilde HO (eds) Ergot alkaloids and related compounds. Springer, New York Berlin Heidelberg, pp 567–614

Goodwin FK (1971) Psychiatric side effects of levodopa in man. JAMA 218: 1915–1920

Hardie RJ, Lees AJ, Stern GM (1984) On-off fluctuations in Parkinson's disease. Brain 107: 487–506

Marsden CD, Parkes JD (1976) "On-off" effects in patients with Parkinson's disease on chronic levodopa therapy. Lancet 1: 292–296

Moskovitz C, Moses H, Klawans HL (1978) Levodopa-induced psychosis: a kindling phenomenon. Am J Psychiatry 135: 669–675

Nutt JG, Woodward WR, Hammerstad JP, Carter JH, Anderson JL (1984) The "on-off" phenomenon in Parkinson's disease. N Engl J Med 310: 483–488

Obeso JA, Martinez-Lage JM, Luquin MR, Bolis N (1983) Intravenous lisuride infusion for Parkinson's disease. Ann Neurol 14: 252

Obeso JA, Luquin MR, Martinez-Lage JM (1986) Lisuride infusion pump: a device for the treatment of motor fluctuations in Parkinson's disease. Lancet 1: 467–470

Quinn NP, Marsden CD, Parkes JD (1982) Complicated response fluctuations in Parkinson's disease: response to intravenous infusion of levodopa. Lancet 2: 412–415

Quinn NP, Parkes JD, Marsden CD (1984) Control of "on-off" phenomenon by continuous intravenous infusion of levodopa. Neurology 34: 1131–1136

Quinn NP, Marsden CD (1986) Lithium for painful dystonia in Parkinson's disease. Lancet 1: 1377

Quinn NP, Critchley PHS, Marsden CD (1987) Young onset parkinsonism. In preparation

Ruggieri S, Stocchi F, Agnoli A (1986) Lisuride infusion pump for Parkinson's disease. Lancet 2: 348–349

Authors' address: P. H. S. Critchley, University Department of Neurology, Institute of Psychiatry, London SE5, U.K.

J Neural Transm (1988) [Suppl] 27: 61–70

A combined regimen of subcutaneous lisuride and oral Madopar HBS in Parkinson's disease

C. Chouza, J. L. Caamaño, Ofrenda de Medina, Ruth Aljanati, A. Scaramelli, and Susana Romero

Section of Extrapyramidal Diseases, Institute of Neurology, Montevideo, Uruguay

Summary. At the first stage of a pilot study involving 14 parkinsonians with motor fluctuations, treatment with standard Madopar was substituted by a sustained-release form, Madopar HBS, which attenuated fluctuations in patients with end-of-dose impairment, but achieved only moderate improvement in patients with on-off phenomena.

In a second phase of the trial, 4 parkinsonians exhibiting the most severe fluctuations of mobility and the poorest response to Madopar HBS (Hydro-dynamically Balanced System) were selected for treatment with a combined regimen utilizing subcutaneous lisuride infusions as the additional component. The sequence of the trial was as follows:

1. standard Madopar,
2. Madopar HBS,
3. standard Madopar combined with lisuride infusions and
4. Madopar HBS combined with lisuride infusions.

Steady improvement was observed along the lines of this schedule, but the best results were obtained when Madopar HBS was combined with lisuride infusions. Subsequently motor fluctuations were less marked or disappeared, early-morning Parkinson symptoms decreased and dystonia was not recorded any longer. Even better results could be accomplished in an extended trial attempting to establish the best dosage ratio of the combination, possibly admitting increased dosage.

The tolerance of the combined regimen was excellent, except in one patient who transiently exhibited delusions and postural hypotension. The combination of sustained-release Madopar and continuous infusions of the dopaminergic agonist lisuride seems to prove a more physiological and effective regimen for the treatment of severe motor fluctuations.

Introduction

Fluctuations of mobility, dyskinesia and dystonia are frequent complications in parkinsonians under chronic levodopa therapy (Chouza et al., 1975; Fahn,

1974; Tolosa et al., 1975), for which both pharmacokinetic and pharmacodynamic factors are to blame (Juncos et al., 1985; Quinn et al., 1984; Shoulson et al., 1975).

For many years investigators have searched for new modes of drug administration or new antiparkinsonian substances with a longer plasma half-life capable of smoothening steep variations of drug plasma levels and, possibly, extinguishing clinical motor oscillations (Juncos et al., 1985). One of the most effective research strategies has been to develop a slow-release preparation of L-dopa plus benserazide, Madopar HBS (100 mg levodopa plus 25 mg benserazide), which achieves satisfactory results in patients with end-of-dose impairment, but is less successful in controlling severe motor fluctuations (Chouza et al., 1987).

However, new modes of drug administration have been developed with a view to provide a sustained delivery system (Obeso et al., 1986 a; Quinn et al., 1984; Shoulson et al., 1975). The subcutaneous infusion of lisuride combined with oral L-dopa treatment has proved to be successful in overcoming severe motor fluctuations (Obeso et al., 1986 b). The combination of a slow-release preparation of L-dopa plus benserazide (Madopar HBS) and subcutaneous infusions of lisuride should achieve even better results considering:

a) The more physiological action of both dopaminergic agents maintaining more sustained plasma levels.

b) Steep changes of the plasma level following the intake of standard Madopar (100 mg levodopa plus 25 mg benserazide) could be avoided, resulting in a more stable therapeutic effect and consequently less marked fluctuations as well as a reduction of dyskinesia and dystonia.

Material and method

The clinical trial was carried out in 2 stages.

First stage

In 14 parkinsonians with motor fluctuations standard Madopar was replaced by Madopar HBS. 10 patients displayed severe and long-standing on-off effects and 4 end-of-dose impairment. Madopar HBS (125 mg) is a sustained-release form employing a "floating capsule" which releases the drug by diffusion after having remained in the stomach for some time. Consequently plasma concentrations are maintained at adequate levels for 6–8 hours following application. The bioavailability of this controlled release formulation, however, is lower than that of standard Madopar necessitating an increased daily dose to achieve comparable therapeutic results.

Second stage

Madopar HBS and subcutaneous infusions of lisuride were used as a combined regimen. Four patients showing the most severe fluctuations and the poorest response to Madopar HBS were selected for this kind of therapy.

The clinical trial followed this stepped-up sequence:

(1) treatment with standard Madopar;

(2) substitution by Madopar HBS;

(3) combination of standard Madopar and subcutaneous infusions of lisuride;

(4) combination of Madopar HBS and subcutaneous infusions of lisuride.

The last 2 steps of the trial took 4–5 weeks for each patient; changes in dose (standard Madopar, Madopar HBS and lisuride) were tested clinically trying to establish the best dose ratio of the combinations on the basis of the results. Subsequently comparative studies were performed between both forms of Madopar in combination with lisuride infusions.

Methodology for the clinical evaluation of motor fluctuations, dyskinesia, and dystonia (Fig. 1)

Continuous daytime assessment (12 h) of each patient involved in the trial allows to evaluate motor performance at intervals of 30 minutes by means of Webster's scale. The numerical addition of the 25 ratings of motor performance collected from each assessment provides the Day Webster Score (DWS).

The degree of early-morning akinesia is derived from the first evaluation prior to the morning application of drugs. The sum total of on-periods, expressed in minutes, contributes another important quantitative reference as to the clinical features, including motor fluctuations and comparisons of the pharmacological effects of drugs applied on these parameters.

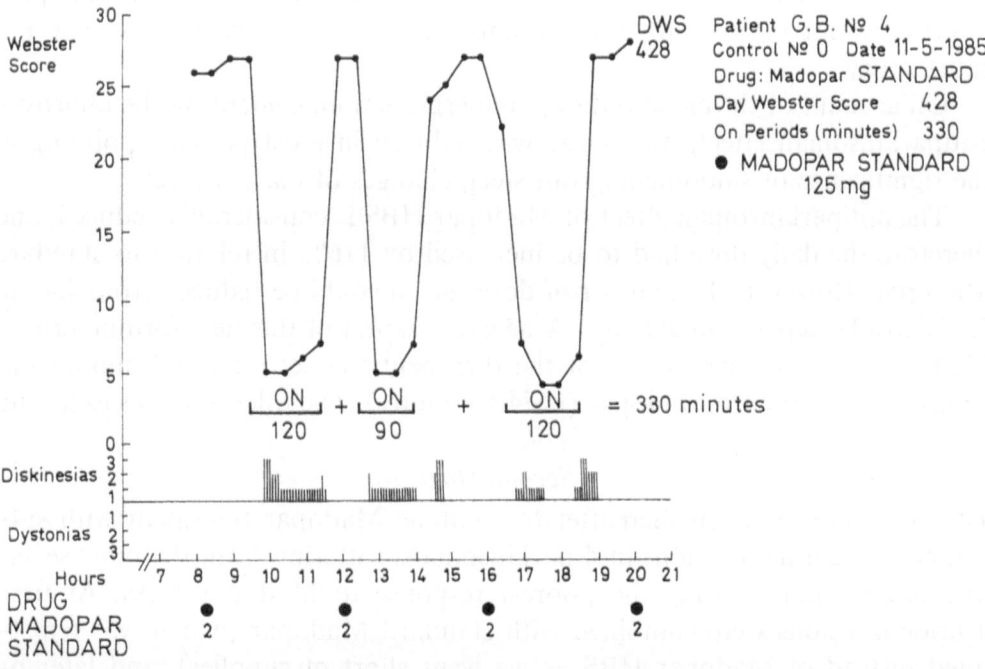

Fig. 1. Clinical assay replacing standard Madopar by Madopar HBS. The symptoms of parkinsonism are evaluated according to the Webster scale at intervals of 30 minutes; the numerical addition of the 25 ratings obtained during 12 hours of continuous clinical observation provides the Day Webster Score (DWS). The overall on-duration during the day results from simply adding on-periods. Day Webster Score and on-duration are the most important reference criteria to assess fluctuations and the antiparkinsonian effect of drugs in this investigation. Dyskinesia and dystonia are recorded according to intensity, duration and frequency

Dyskinesia and dystonia were registered according to type, intensity, frequency and duration. Self-rating records were filled in every day by the patients and their relatives. Blood pressure and pulse rate were recorded as well in order to trace early postural hypotension.

Results

First stage

Madopar HBS has proved to have distinct advantages over standard Madopar for the following reasons:

a) sustained therapeutic effect extending on-periods;

b) reduction or disappearance of fluctuations with less abrupt changes from on to off and vice versa;

c) off-periods are less marked and consequently motor performance is improved at this stage;

d) improvement of nocturnal pain and motor behaviour as well as of early-morning akinesia. Early-morning symptoms disappeared in 4 patients, a marked decrease was observed in 2 and a mild reduction was registered in some others;

e) these advantages were more distinct in patients with end-of-dose impairment than in those showing on-off phenomena. Fluctuations disappeared in the 4 patients with end-of-dose impairment. In patients with on-off phenomena motor oscillations were diminished, but the global results were unsatisfactory.

On account of the enhanced dopaminergic action, evident by the improved antiparkinsonian effect, dyskinesia was reduced in most patients pointing at the significance of smoothening out steep changes of plasma levels.

The antiparkinsonian effect of Madopar HBS is considerably reduced, and therefore the daily dose had to be increased by 116% in relation to standard Madopar. However, the number of doses given could be reduced from 4.43 to 3.57 capsules/day on an average. A negative aspect of this new formulation of Madopar is the delayed onset of the therapeutic effect ($\bar{x}$: 120.5 minutes) as compared to standard Madopar ($\bar{x}$: 74.5 minutes). The tolerance was excellent.

Second stage

The decision was taken thereafter to combine Madopar treatment with subcutaneous lisuride infusions in 4 parkinsonians suffering from the most severe fluctuations and showing the poorest response to Madopar HBS. At first, lisuride infusions were combined with standard Madopar (which was reinstituted instead of Madopar HBS as we went short of supplies), and later on standard Madopar was exchanged by Madopar HBS.

The changes induced by standard Madopar, Madopar HBS, standard Madopar + lisuride infusion, and Madopar HBS + lisuride infusion, recorded by the Day Webster Score (DWS), on-periods, early-morning symptoms and the alterations of dyskinesia and dystonia, were taken as a reference to arrive at conclusions about the pharmacological effects of these drug combinations (Figs. 2–5).

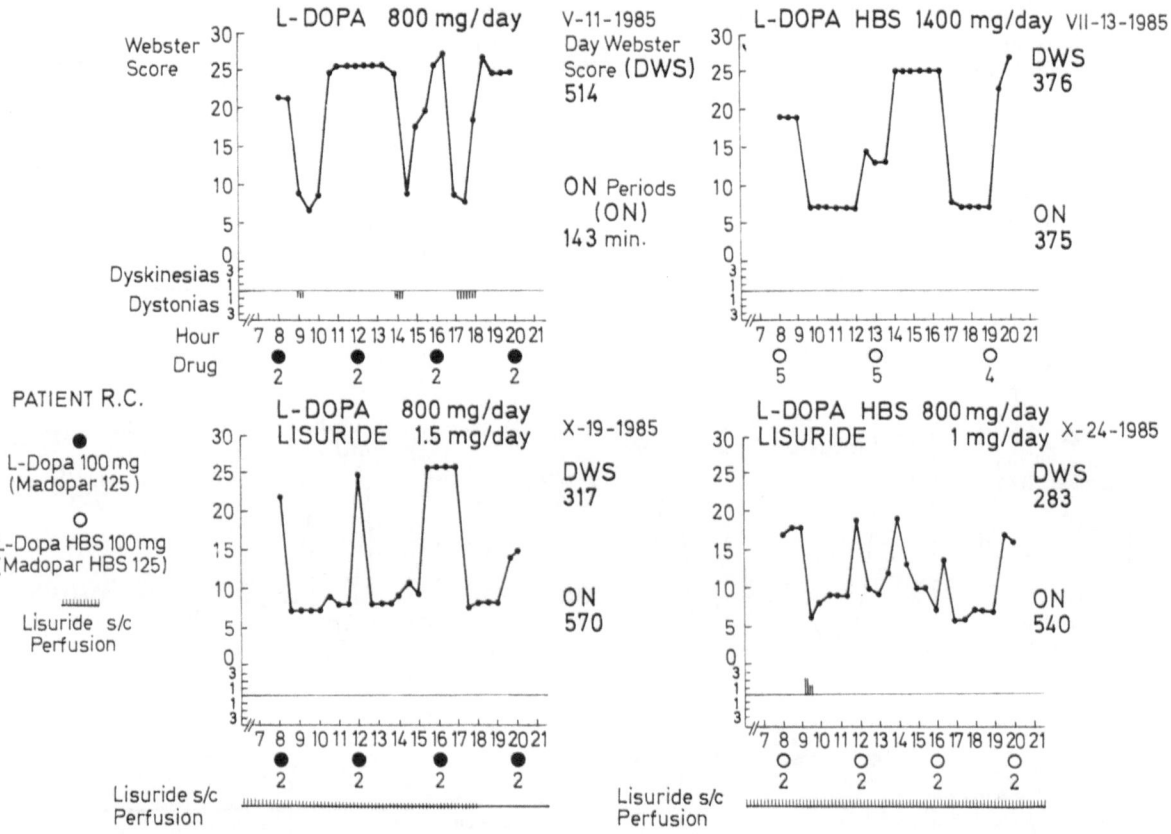

Fig. 2. Treatment of the on-off effect with Madopar HBS and subcutaneous infusion of lisuride. Patient no. 1 R. C.: The best therapeutic result was obtained with the combination of Madopar HBS and lisuride infusions. Fluctuations were less pronounced and dystonia disappeared. This improvement was achieved with low doses of both drugs: 800 mg of Madopar HBS have a similar therapeutic effect as compared to 400 mg or less of standard Madopar; 1 mg of lisuride is also considered to be a low dose which could not be increased as the patient exhibited delusions which disappeared following dose reduction. He also suffered from postural hypotension which was controlled by domperidone 80 mg/day

The ratings of the Day Webster Score and on-duration with corresponding improvement at each stage of treatment are visualized in Figs. 2–5. It is clearly evident that the progressive improvement reaches its optimum when using the combination of Madopar HBS and lisuride infusions.

Tolerance

Tolerance of Madopar HBS was excellent. Lisuride infusions were also well tolerated, except in case 1 who exhibited delusions when receiving 1.5 mg/day, but disappearing with the reduction of dose to 1 mg/day. In addition, this patient suffered from postural hypotension which was overcome by increasing the dose of domperidone up to 80 mg/day.

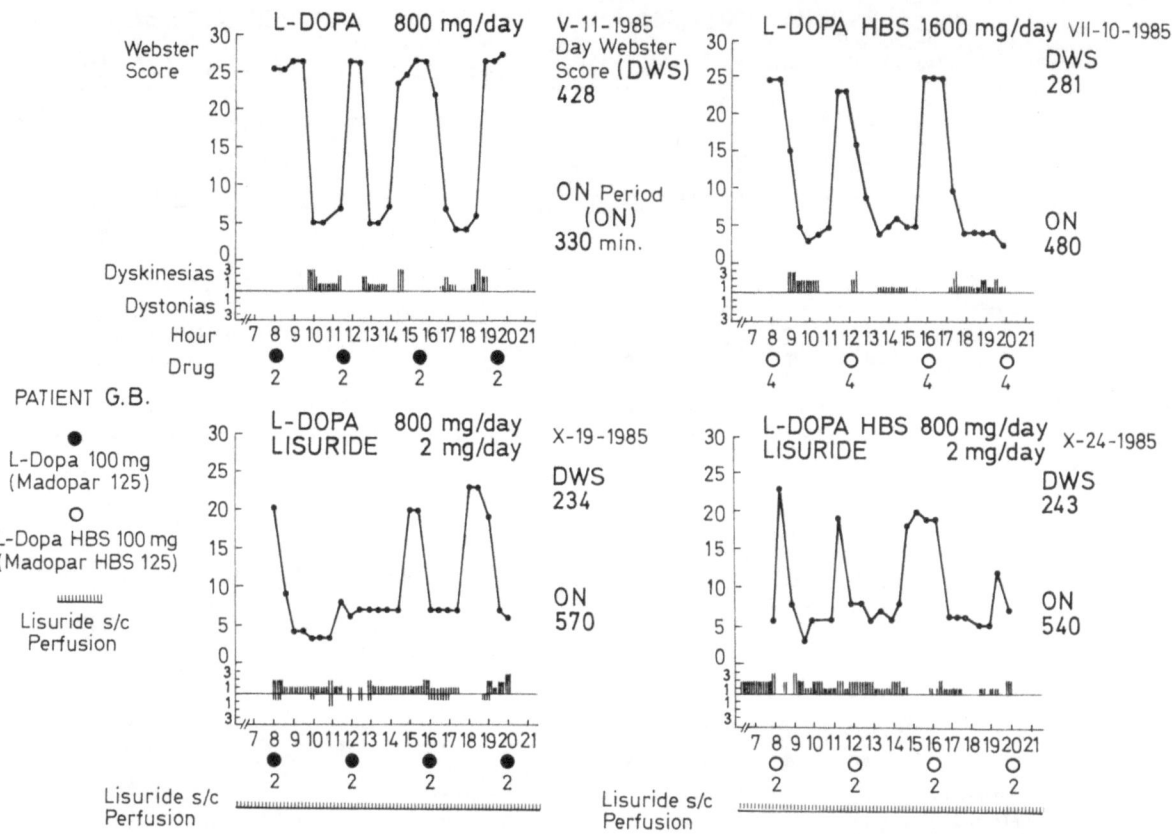

Fig. 3. Treatment of the on-off effect with Madopar HBS and subcutaneous infusion of lisuride. Patient no. 2 G. B.: The benefits of the stepped-up regimen evidenced by reduced fluctuations, as expressed in amplitude and frequency, are obvious. The combination of Madopar HBS and lisuride achieved the lowest ratings of early-morning symptoms proving also that the incidence of nocturnal symptoms was negligible

Dosage

The minimal and maximal doses of Madopar and lisuride, when used in combination, were as follows:

standard Madopar	500–800 mg/day
Madopar HBS	800–1200 mg/day
Lisuride	1–2 mg/day

Employing lisuride infusions as an additional regimen allows to curb the daily dose of L-dopa by 6.9% as applied with standard Madopar and 33.3% as with Madopar HBS. This difference, to a large extent, is explained by the lower bioavailability of L-dopa used in Madopar HBS and the increased dosage required therefore.

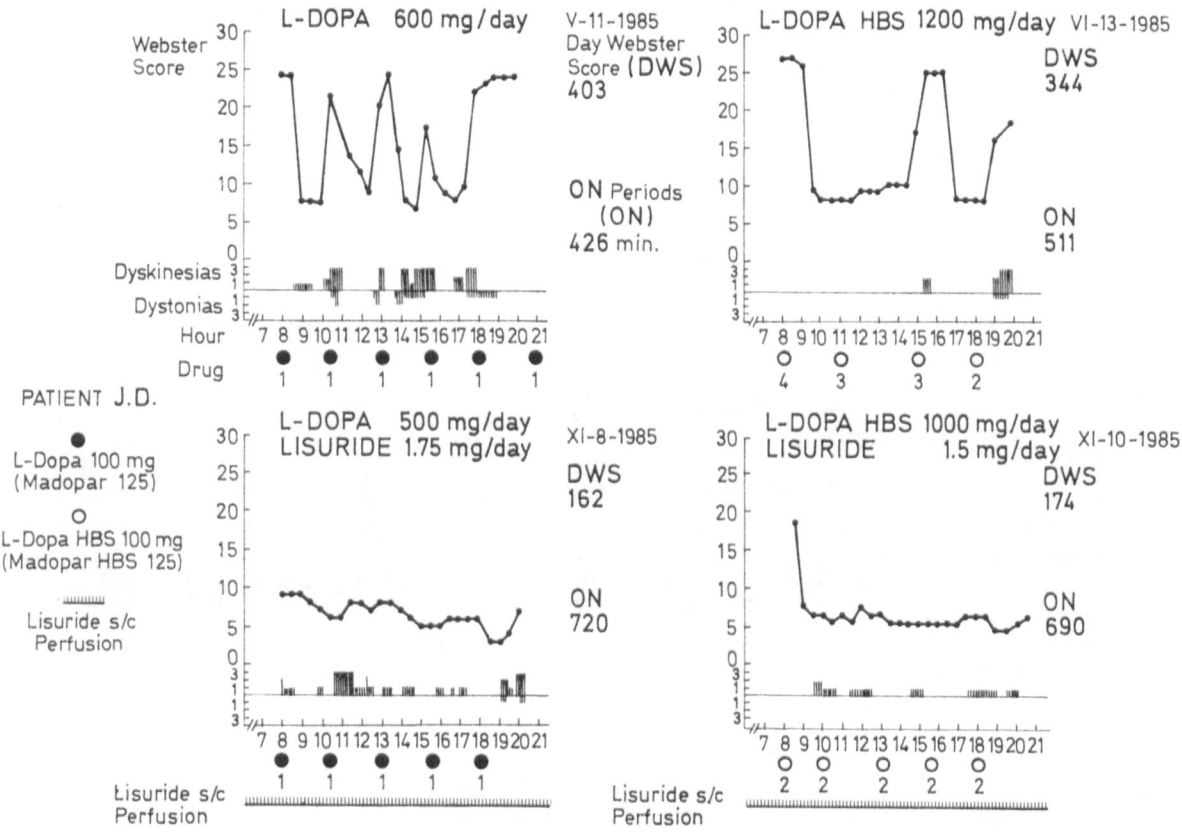

Fig. 4. Treatment of the on-off effect with Madopar HBS and subcutaneous infusion of lisuride. Patient no. 3 J. D.: The first graph demonstrates the typical course of fluctuations marked by on-off phenomena with steep motor changes and extensive dyskinesia as well as dystonia. The substitution of standard Madopar by Madopar HBS led to clear improvement of all symptoms. The combination with lisuride infusions smoothened the curve and gave it practically a linear shape throughout the day with only minimal symptoms occurring. The Day Webster Score (DWS) changed from 403 to 162 and 174 in the last two controls under lisuride infusion treatment. Dyskinesia was markedly reduced due to the enhanced dopaminergic effect, and dystonia disappeared. The tolerance of the regimen was excellent

Discussion

This pilot study suggests distinct therapeutic advantages combining a sustained-release form of Madopar with the continuous s.c. infusion of lisuride. The benefits of subcutaneous lisuride infusions in the treatment of motor fluctuations in chronic levodopa therapy are achieved by continuous dopaminergic stimulation. This procedure, originally applied by Obeso et al. (1986 b), unfortunately could not avoid steep plasma changes prevailing under standard Madopar. The combination of slow-release Madopar, however, was supposed to overcome this problem and, in fact, this assumption has been confirmed in our clinical trial.

68 C. Chouza et al.

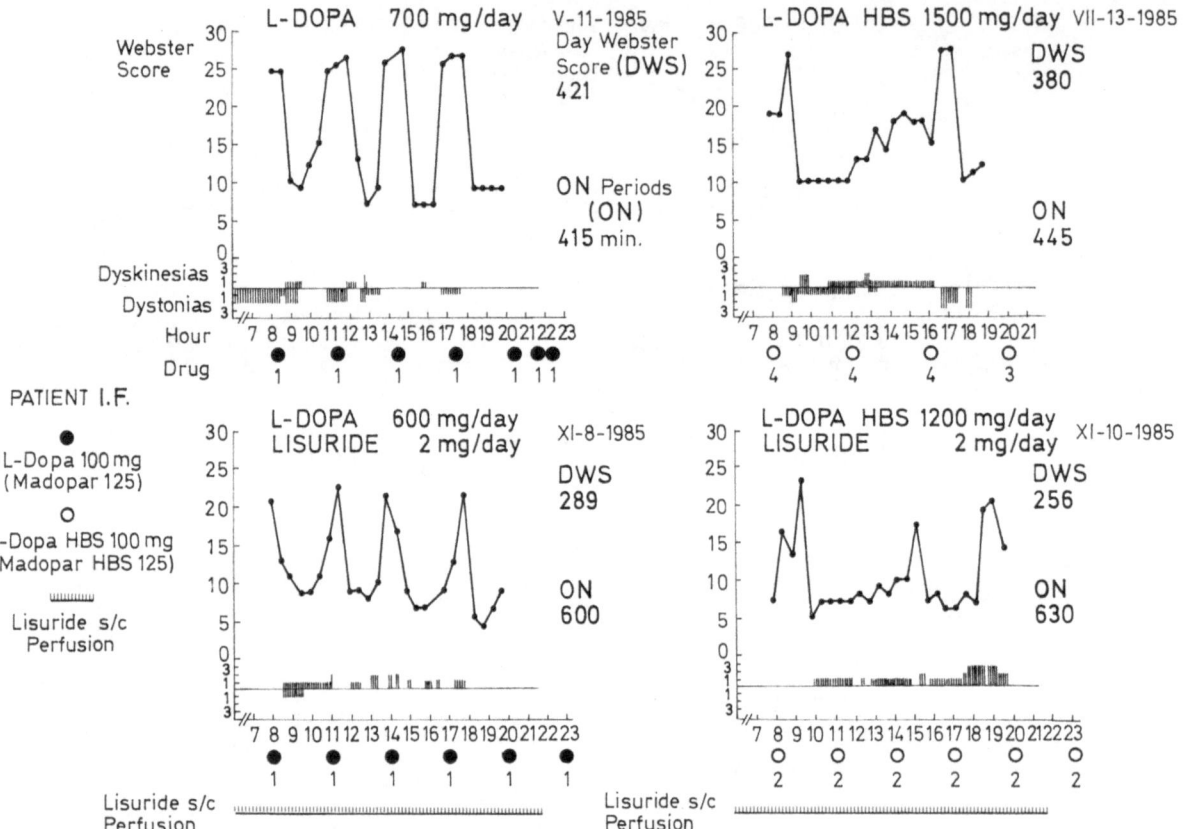

Fig. 5. Treatment of the on-off effect with Madopar HBS and subcutaneous infusion of lisuride. Patient no. 4 I. F.: The first control, recorded under treatment with standard Madopar, shows on- and off-periods of an almost geometrical appearance, typically associated with severe long-lasting nocturnal and morning dystonia as well as often generalized dystonia. The substitution by Madopar HBS resulted in a global improvement except for a mild increase of dyskinesia, but a reduction of dystonia. The combination of standard Madopar and lisuride infusions could not avoid the occurrence of off-periods which were markedly reduced under Madopar HBS. Finally, in the fourth stage of the trial the combination of Madopar HBS and lisuride infusions produced the best therapeutic results

Steady improvement has resulted from the following stepped-up regimen:
(1) standard Madopar,
(2) Madopar HBS,
(3) standard Madopar plus subcutaneous infusions of lisuride,
(4) Madopar HBS plus subcutaneous infusions of lisuride.

The most significant level of improvement was reached by the addition of subcutaneous infusions of lisuride. Together with Madopar HBS this combination achieved the best results, after all.

Other beneficial effects of this combination were the attenuation or even disappearance of motor fluctuations and reduction of early-morning symptoms. Disappearance of dystonia in 4 cases treated with the combination of Madopar HBS and lisuride infusions provides evidence for the assumption that a more extensive and sustained dopaminergic effect had been exerted. Dyskinesia may also be diminished by this combination, as is demonstrated in case No. 3 (Fig. 4). On the contrary, cases 2 and 4 exhibited an increased incidence of dyskinesia which seemed to correlate with the maximum level of plasma L-dopa under Madopar HBS prevalent in the evening. Better results possibly could not be achieved due to the low doses of both drugs employed, especially lisuride, and the short duration of the clinical trial in addition to insufficient time for establishing an optimal ratio of the drug combination. The tolerance was excellent in 3 patients treated. Only in one patient psychiatric symptoms became apparent, but disappeared concurrent with the dose reduction of lisuride. Postural hypotension was controlled in this patient by increasing the dose of domperidone.

Our method of continuous observation used together with a quantitative evaluation of motor performance allows for a more objective overall clinical assessment of the complex on-off phenomenon. It is well realized that fluctuations are not a schematic representation of identical everyday features. Their wide range of variations requires a multiple number of clinical observations to arrive at subtle conclusions. The observations under discussion were not made in isolation but represent the tendency of clinical events prominent under each drug combination.

The combination of L-dopa plus decarboxylase inhibitors (DI) and dopaminergic agonists is considered the best treatment for patients with Parkinson's disease. The administration of slow-release L-dopa plus benserazide (Madopar HBS) and continuous infusions of the dopaminergic agonist lisuride provides more sustained levels of both dopaminergic agents and better therapeutic results, accordingly. We believe that a combination of these two new drug delivery systems procures a more physiological and effective treatment in Parkinson's disease, forcefully reducing severe fluctuations common in chronic levodopa therapy.

Acknowledgement

We wish to thank Hoffmann-La Roche Laboratories for the supply of Madopar HBS; Schering AG for providing lisuride; Dr. José Obeso and his colleagues for their contributions of material and advice on the lisuride subcutaneous infusion system.

References

Chouza C, Romero S, Gomensoro JB (1975) "Long-term syndrome" en el tratamiento del parkinsonismo con L-dopa. Acta Neurol Latinoam 21: 108–125

Chouza C, Romero S, de Medina O, Aljanati R, Scaramelli A, Caamaño JL, Gonzalez Panizza V (1987) Substitution of standard Madopar for Madopar HBS in parkinsonians with fluctuations. Eur Neurol 27 [Suppl 1]: 59–67

Fahn S (1974) On-off phenomenon with levodopa therapy in parkinsonism. Neurology 24: 431–441

Juncos J, Serrati C, Fabbrini G, Chase TN (1985) Fluctuating levodopa concentrations and Parkinson's disease. Lancet 1: 440

Obeso JA, Luquin MR, Martinez-Lage JM (1986) Intravenous lisuride corrects oscillations of motor performance in Parkinson's disease. Ann Neurol 19: 31–35

Obeso JA, Luquin MR, Martinez-Lage JM (1986) Lisuride infusion pump: a device for the treatment of motor fluctuations in Parkinson's disease. Lancet 1: 467–470

Quinn N, Parkes P, Marsden D (1984) Control of on-off phenomenon by continuous intravenous infusion of levodopa. Neurology 34: 1131–1136

Shoulson I, Glaubiger GA, Chase TN (1975) On-off response: clinical and biochemical correlations during oral and intravenous levodopa administration in parkinsonian patients. Neurology 25: 1144–1148

Tolosa ES, Martin WE, Cohen HP, Jacobson RL (1975) Patterns of clinical response and plasma dopa levels in Parkinson's disease. Neurology 25: 177–184

Authors' address: Dr. C. Chouza, Instituto de Neurología Hospital de Clínicas, Montevideo, Uruguay.

J Neural Transm (1988) [Suppl] 27: 71–74

Pharmacokinetics of lisuride after subcutaneous infusion

W. Krause[1], **B. Nieuweboer**[1], **St. Ruggieri**[2], **F. Stocchi**[2], and **I. Suchy**[1]

[1] Research Laboratories of Schering, Berlin and Bergkamen,
Federal Republic of Germany
[2] Department of Neurological Science, University of Rome "La Sapienza", Rome, Italy

Summary. Six parkinsonian patients received a constant subcutaneous infusion of 60 µg lisuride per hour in the abdominal region for 2 hours. Plasma levels of the unchanged drug were measured by radio-immunoassay. During infusion, a steady state plasma level of 0.78 ± 0.19 ng/ml was achieved. After discontinuation of the infusion, concentrations declined with a half-life of 1.4 ± 0.4 hour. The total clearance of lisuride was 20 ± 6 ml/min/kg. Due to the low interpatient variability of plasma levels, a good control of clinical effects is to be expected.

*

As with many other ergot derivatives, pharmacokinetic studies with lisuride in humans have been rather difficult. This, amongst other factors, is due to the very low therapeutic plasma levels, high intra- and especially interindividual variability and the chemical and metabolic instability of the compound in biological samples. To be able to assess the low plasma levels of lisuride in clinical studies, we developed a highly specific and sensitive radioimmunoassay with a detection limit of approx. 20 pg/ml, which is far below the therapeutic concentrations (Hümpel et al., 1981). This RIA was used in all studies in volunteers and patients described below.

Table 1. Pharmacokinetic parameters of lisuride in six volunteers after oral administration (from Hümpel et al., 1981, 1984)

Absorption	almost complete
Biovailability	$15 \pm 6\%$ of oral dose
C_{max} (0.3 mg p.o.)	280 ± 80 pg/ml
t_{max}	1.2 ± 1.0 h
Terminal half-life	1.9 ± 0.6 h
Total clearance	13 ± 4 ml/min $\times$ kg
Plasma protein binding	approx. 65%

In humans, lisuride is completeley metabolized and the biotransformation products are excreted to nearly equal proportions with urine and faeces. In agreement with its short half-life, no accumulation of lisuride was observed on repeated oral administration.

In parkinsonian patients comparable pharmacokinetic data for lisuride have been described (Burns et al., 1984). Also, on long-term administration in these patients, individual bioavailability and other kinetic parameters remained rather constant (Rinne and Hümpel, in preparation).

As a consequence of the low and variable bioavailability, parenteral treatment was investigated as well (Hümpel et al., 1981; Burns et al., 1984). Figure 1 shows lisuride plasma levels observed after oral, intravenous and intramuscular administration of a single dose of lisuride to healthy volunteers (data from Hümpel et al., 1981, and Dorow, unpublished). As in the clinical studies (Dorow et al., 1980), lisuride was dissolved in saline when administered parenterally. As could be expected from the time course of the plasma levels (with the exception of the prolactin-lowering effect) clinical effects after a single parenteral dose were shortlasting (Hümpel et al., 1981; Suchy and Horowski, 1984). Therefore,

Table 2

Patient	B	C	D	E	F	G
Weight (kg)	50	59	85	56	76	56
Height (cm)	155	159	165	155	169	165
Sex	F	F	F	F	M	M

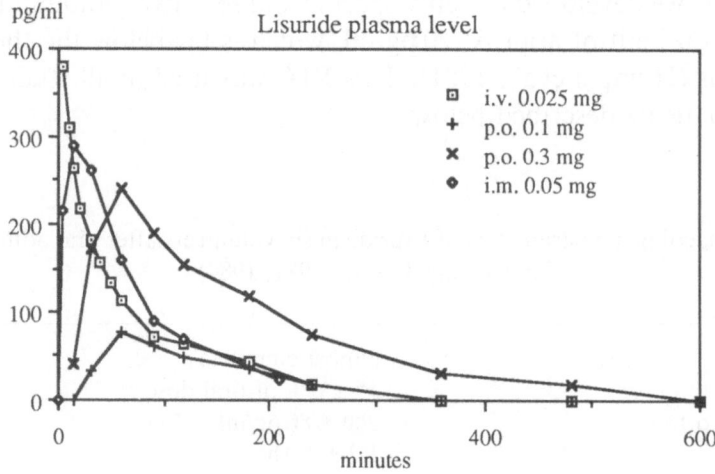

Fig. 1. Mean plasma levels of lisuride in six volunteers after i.v. injection of 0.025 mg and p.o. administration of 0.1 and 0.3 mg (Hümpel et al., 1981) and in three volunteers after i.m. injection of 0.05 mg (Dorow, unpublished)

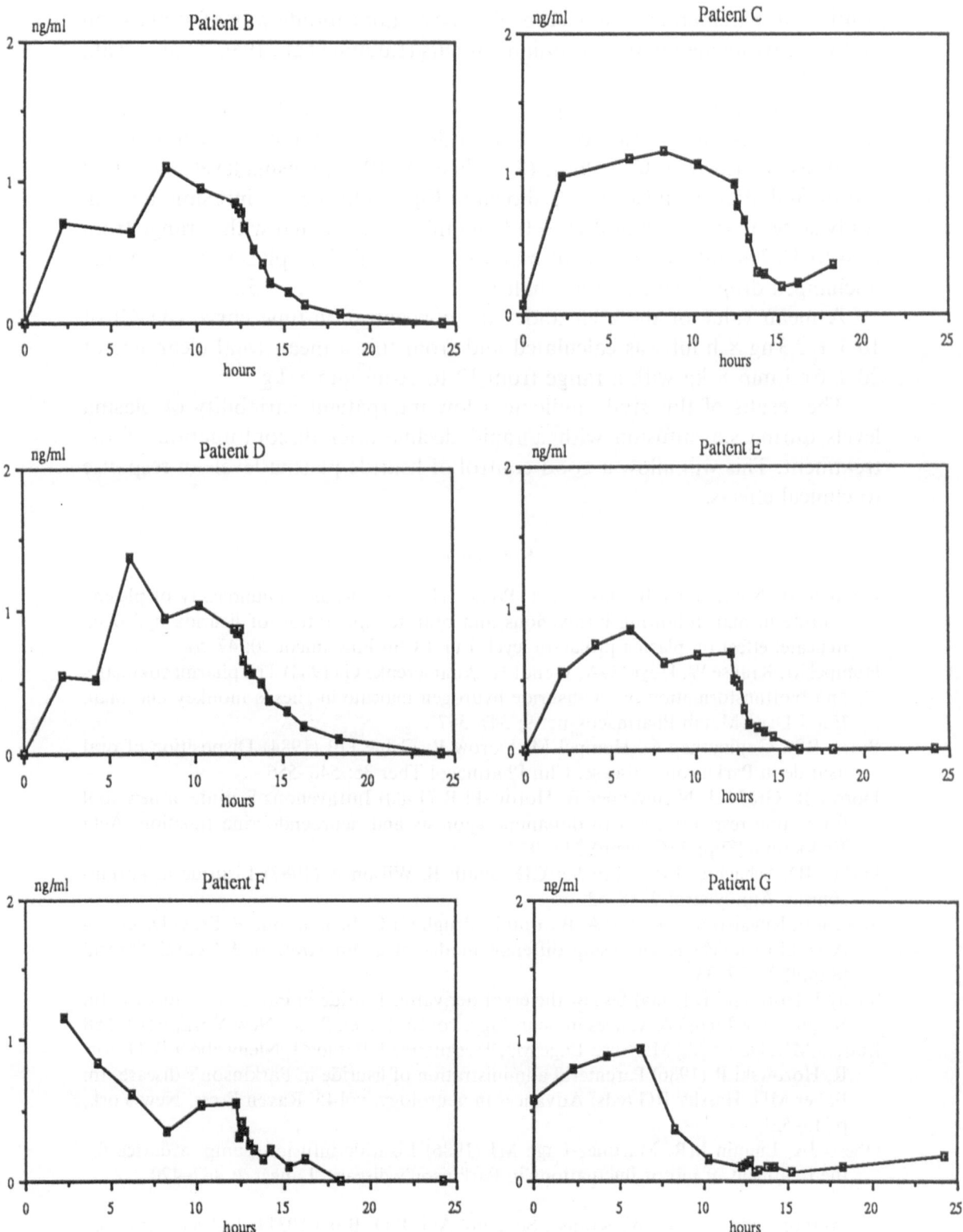

Fig. 2. Plasma levels of lisuride in parkinsonian patients during and after subcutaneous
infusion of 60 μg/h for 12 hour

continuous parenteral infusion seemed to be an appropriate way of application and was investigated in parkinsonian patients (Luquin et al., 1986; Obeso et al., 1986).

In a recent clinical study, 6 parkinsonian patients (Table 2) received a constant subcutaneous infusion of 60 µg lisuride/h in the abdominal region for 12 hours, as described in this volume (Stocchi et al., 1988). Plasma levels measured during and after the infusion are shown in Fig. 2. During s.c. infusion, a mean study state plasma level of 0.78 ± 0.19 ng/ml was evaluated with a range from 0.54 to 1.07 ng/ml. After discontinuation of the infusion, plasma levels of the unchanged drug declined with a half-life of 1.4 ± 0.4 h (n = 5).

A mean value of the area under the concentration-time curve (AUC) of 10.4 ± 2.3 ng × h/ml was calculated and from this a mean total clearance of 20 ± 6 ml/min × kg with a range from 12 to 26 ml/min × kg.

The results of this study indicate a low interpatient variability of plasma levels during s.c. infusion with a rapid decline after discontinuation of the treatment. This will allow a good control of lisuride plasma levels in response to clinical effects.

References

Hümpel M, Nieuweboer B, Hasan SH, Wendt H (1981) Radioimmunoassay of plasma lisuride in man following intravenous and oral administration of lisuride hydrogen maleate; effect on plasma prolactin level. Eur J Clin Pharamcol 20: 47–51

Hümpel M, Krause W, Hoyer GA, Wendt H, Pommerenke G (1984) The pharmacokinetics and biotransformation of ^{14}C-lisuride hydrogen maleate in rhesus monkey and man. Eur J Drug Metab Pharmacokinet 9: 347–357

Burns RSt, Gopinathan G, Hümpel M, Dorow R, Calne DB (1984) Disposition of oral lisuride in Parkinson's disease. Clin Pharmacol Ther 35: 548–556

Dorow R, Gräf KJ, Nieuweboer B, Horowski R (1980) Intravenous lisuride: a new tool for testing responsiveness to dopamine agonists and neuroendocrine function. Acta Endocrinol [Suppl] (Copenh) 234: 9

Parkes JD, Schachter MB, Marsden CD, Smith B, Wilson A (1987) Lisuride in Parkinsonism. Ann Neurol 9: 48–52

Stocchi F, Ruggieri S, Antonini A, Baronti F, Brughitta G, Bellantuono P, Bravi D, Agnoli A (1988) Clinical results using different modes of administration. J Neural Transm [Suppl] 27: 27–33

Suchy I, Horowski R (1984) Use of the ergot derivative lisuride in Parkinson's disease. In: Bergmann KJ (eds) Advances in neurology, vol 40. Raven Press, New York, p 561–568

Luquin MR, Obeso JA, Martinez-Lage MJ, Tresguerres J, Parada J, Nieuweboer B, Dorow R, Horowski R (1986) Parenteral administration of lisuride in Parkinson's disease. In: Baker MD, Hassler RG (eds) Advances in neurology, vol 45. Raven Press, New York, p 515–521

Obeso JA, Luquin MR, Martinez-Lage MJ (1986) Lisuride infusion pump: a device for the treatment of motor fluctuations in Parkinson's disease. Lancet 2: 467–470

Authors' address: Dr. W. Krause, Schering AG, P.O. Box 650311, D-1000 Berlin 65.

J Neural Transm (1988) [Suppl] 27: 75–84

Treatment of Parkinson's disease with subcutaneous lisuride infusions

M. Fernandez Pardal, F. Micheli, Mabel Gatto, and **Natividad Perez y Gonzalez**

Department of Neurology, Hospital de Clínicas José de San Martín,
University of Buenos Aires, Buenos Aires, Argentina

Summary. Four patients with Parkinson's disease and severe fluctuating responses to levodopa and oral dopamine agonists were treated with continuous administration of lisuride infusions, administered by means of an externally worn pump.

Levodopa dosage ranged from 300 to 687 mg/day and was kept stable throughout the study.

In addition increasing doses of lisuride were injected subcutaneously in the abdomen.

Lisuride doses ranged from 41 to 104 µg/h.

A marked improvement in mobility was observed in every patient while severe biphasic dyskinesais almost remitted in one of them.

The most common side-effect was the presence of subcutaneous nodules appearing at the injection site.

Two cases had mild hemorrhagic complications and one initially had nausea. One patient developed acute psychiatric disturbances severe enough to be excluded from the study.

Our findings suggest that lisuride subcutaneous infusions can be useful in severily handicapped parkinsonian patients, however local and psychiatric side-effects may be a serious threat in the long-term care.

Introduction

After the first years of treatment as many as 50% of parkinsonian patients, who initially had a smooth response to levodopa treatment, develop fluctuations in motor performance (Cotzias et al., 1967; Yahr et al., 1969; Rinne et al., 1970; Barbeau, 1974; Presthus and Holmsen, 1974; Marsden and Parkes, 1976). Predictable fluctuations in disability are mainly caused by oscillations in levodopa plasma levels while those occurring in a random fashion have been ascribed to pharmacodynamic factors (Fahn, 1974; Calne et al., 1974; Tolosa et al., 1975; Lhermitte et al., 1977; Perret et al., 1977; Marsden et al., 1982; Garcia de Yébenes et al., 1982; Gervás et al., 1983; Nutt et al., 1984; Hardie et al., 1984). Nevertheless, frequent clinical and levodopa monitoring often discloses a discernible pattern even in otherwise random fluctuators (Nutt and Woodward, 1983). Intravenous levodopa infusions have repeatedly proven capable of producing a sustained response in parkinsonian patients who exhibit unresponsive oscillations (Shoulson et al., 1975; Marsden, 1977; Glough, 1982; Gervás et al., 1983, Quinn et al., 1984).

Recently Obeso et al. demonstrated that continuous administration of lisuride either subcutaneously or intravenously had similar effects (Obeso et al., 1983 a, b; 1986). Subcutaneous administration has so far proven more suitable than intravenous treatment.

Here we report our experience using subcutaneous lisuride (SCL) infusions in four parkinsonian patients.

Patients and methods

Subjects

Four patients (3 women and one man) with idiopathic Parkinson's disease (PD) and fluctuating response to levodopa and oral dopamine agonists (bromocriptine or lisuride), participated in the trial.

Ages ranged from 57 to 75 years with an average of 64, and the duration of symptoms varied from 9 to 17 years with a mean of 12.2 years.

All of them were in stage IV of Hoehn and Yahr and were predictable responders. Three had end of dose deterioration and 3 had drug resistant Off periods. In addition all had biphasic and peak of dose dyskinesias, one had early morning dystonia and one Off dystonia.

The duration of levodopa therapy varied from 8 to 16 years with a mean of 11.5 years.

At the initiation of the study levodopa dosage ranged from 300 to 687 mg/day and was combined with either benserazide or carbidopa. Three patients also received bromocriptine and one oral lisuride.

None had a history of psychiatric disturbances although one (case 2) had an abnormal Mini Mental Status Examination score (24/30).

Clinical features of each patient are summarized in Table 1.

Methods

Patients were admitted to the Neurology Department of the Hospital de Clínicas and consented to the trial according to the Helsinki declaration.

They were evaluated hourly using the King's College Hospital Parkinson's Disease Rating Scale for a baseline period of 3 days, and also during their stay in the Hospital.

Sequential and cumulative On-Off charts were recorded over several days before and during the SCL infusions, given by means of a programmable pump (Tecensa-aip 62, Madrid, Spain) (Figs. 1, 2, and 3).

Dyskinesias were evaluated taking into account the degree of interference with voluntary movements, their distribution and type, as well as the relation to levodopa intake (Figs. 4, 5, and 6).

Besides, the patients were asked to keep a full On-Off diary on a hourly basis. Oral domperidone 30 mg/day was given in 3 fractional doses and 3 days later dopamine agonists were discontinued and SCL infusions started.

Levodopa dosage was kept stable except in patient 1, in whom it was transiently halved during 9 days.

Table 1. Clinical characteristics of the patients

Patient/Age	Years since diagnosis	Years on levodopa therapy	Stage of disease	Disability score		Type of motor fluctuation	Dyskinesias	Treatment when studied	
				On	Off			L-DOPA + INH mg/day	Bromo-criptine mg/day
1/56	17	16	IV	42	69	complicated end of dose with drug resistant off periods	square wave, biphasic, benefit of dose	500 + 50 (c)	6.25
2/66	9	8	IV	28	66	end of dose deterioration with drug resistant off periods	early morning dystonia, biphasic, benefit of dose	687 + 68 (c)	10
3/75	12	11	IV	13	54	complicated end of dose with drug resistant off periods	biphasic, benefit of dose, off period dystonia	300 + 75 (c)	oral lisuride 1.6 mg
4/75	11	11	IV	25	80	complicated end of dose deterioration	biphasic, benefit of dose	650 + 162 (d)	10

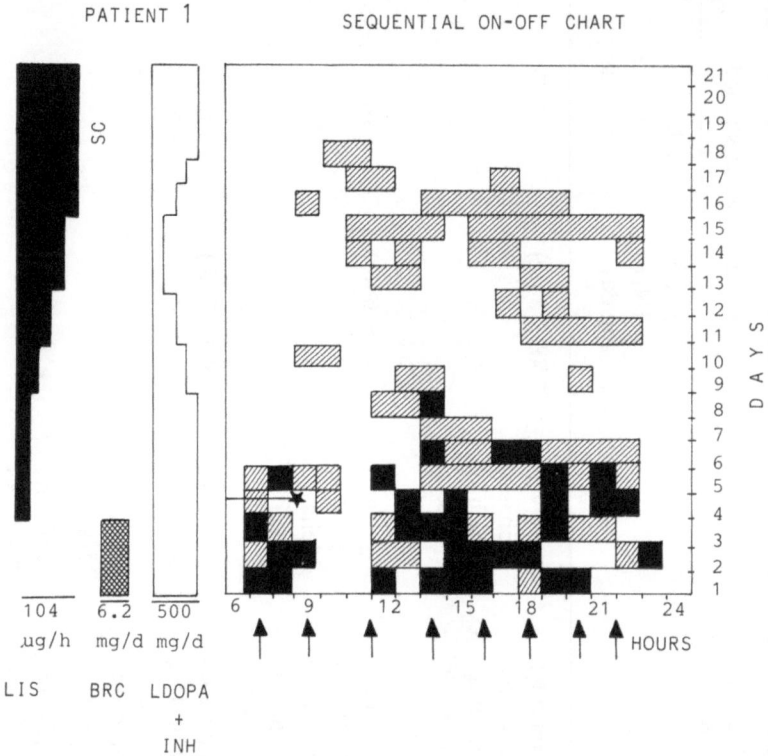

Fig. 1. ■ Immobility; ▨ reduced mobility; □ good mobility; ★ start of SCL infusions; ↑ L-dopa doses; *BRC* bromocriptine

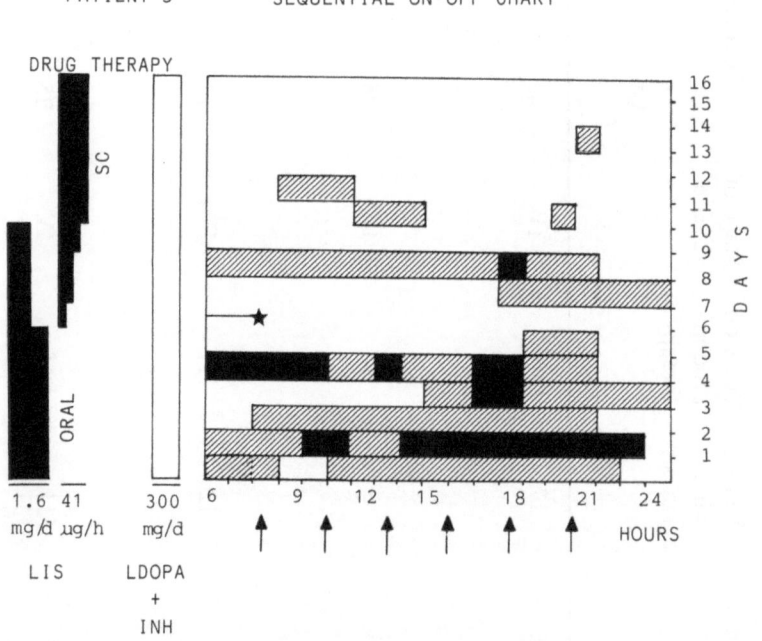

Fig. 2. ■ Immobility; ▨ reduced mobility; □ good mobility; ★ start of SCL infusions; ↑ L-dopa doses; *BRC* bromocriptine

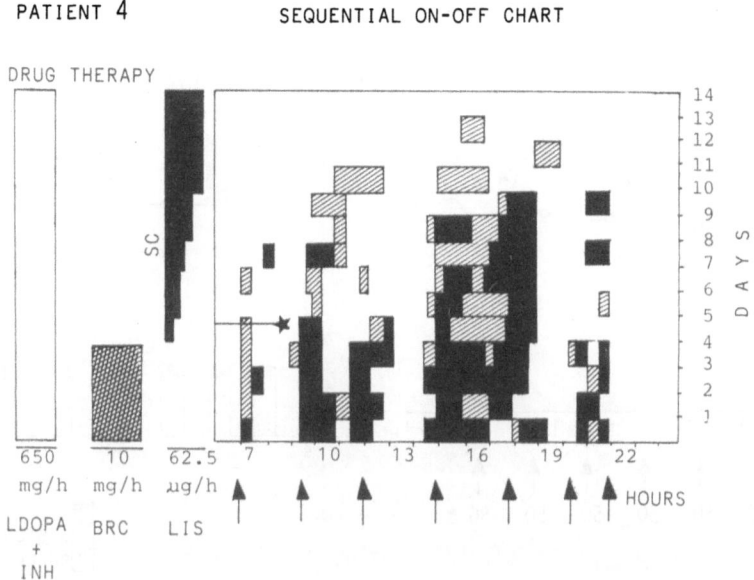

Fig. 3. ■ Immobility; ▨ reduced mobility; ☐ good mobility; ★ start of SCL infusions; ↑ L-dopa doses; *BRC* bromocriptine

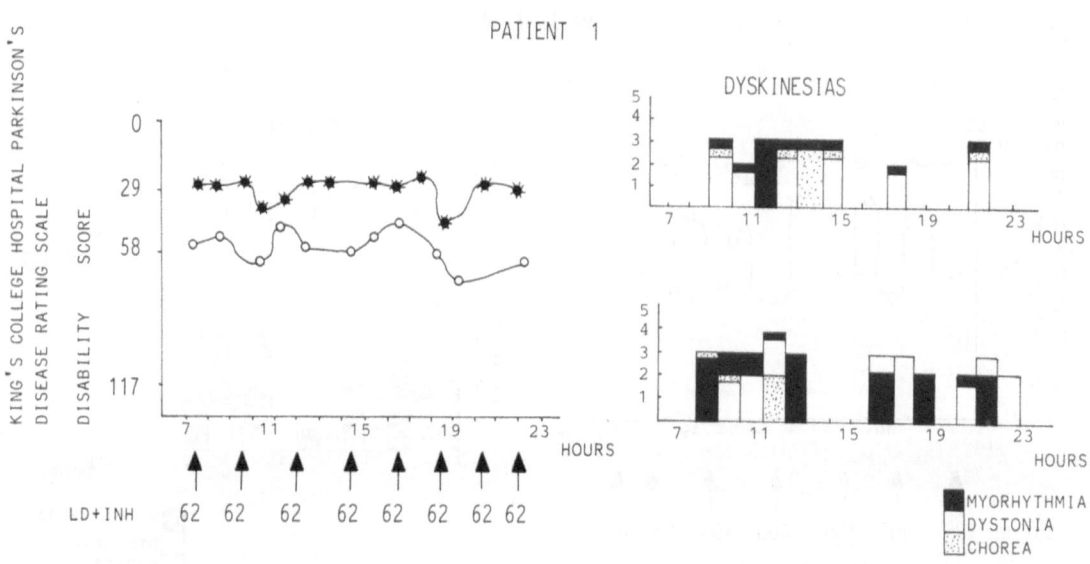

Fig. 4. Mean disability score (o—o) over the 3 days before starting treatment with SCL. Response to SCL infusions (104 µg/h) during the second week of treatment (*—*). Functional incapacity caused during dyskinesias graded from 1 (no interference with any voluntary movements) to 5 (severe Interference). Top: Lisuride sc 104 µg/h. Bottom: Baseline period

PATIENT 3

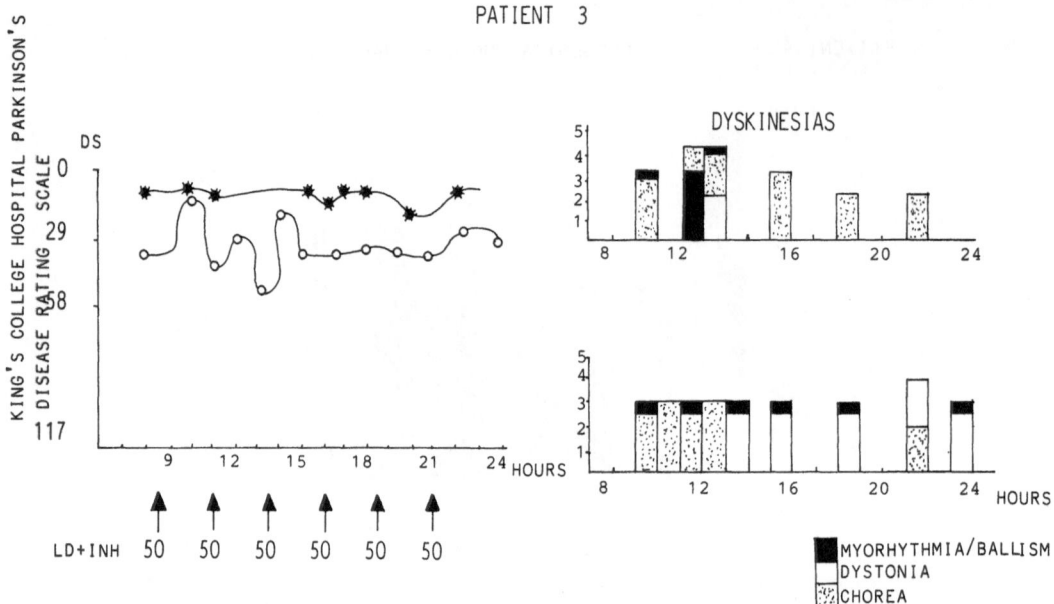

Fig. 5. Mean disability score (o—o) over the 6 days before starting treatment with SCL. Response to SCL infusions (41.6 μg/h) during the second week of treatment (*—*). Functional incapacity caused during dyskinesias graded from 1 (no interference with any voluntary movements) to 5 (severe interference). Top: Lisuride sc 41.6 μg/h. Bottom: Baseline period

PATIENT 4

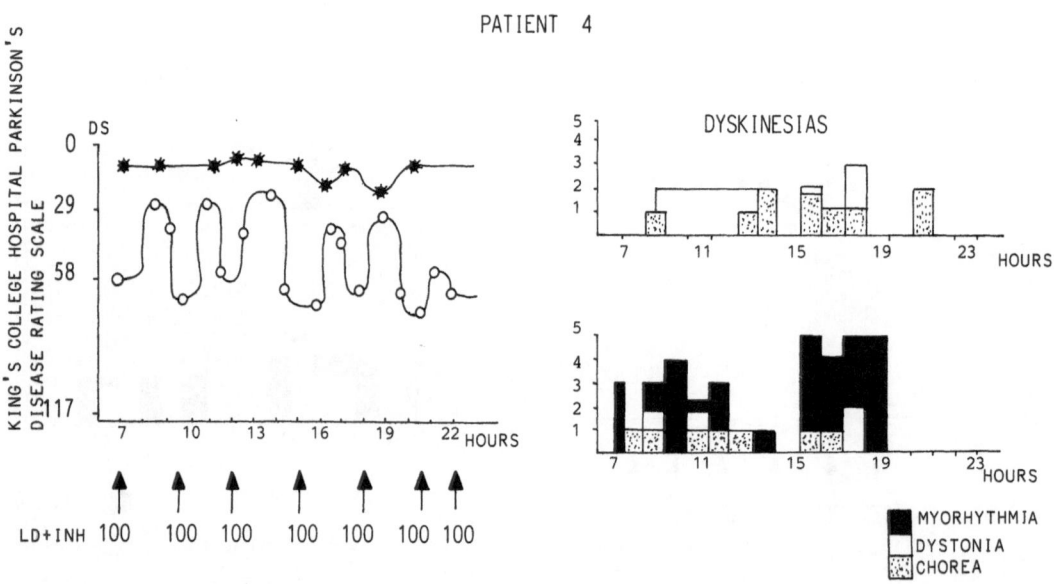

Fig. 6. Mean disability score (o—o) over the 4 days before starting treatment with SCL. Response to SCL infusions (62.5 μg/h) during the second week of treatment (*—*). Functional incapacity caused during dyskinesias graded from 1 (no interference with any voluntary movements) to 5 (severe interference). Top: Lisuride sc 62.5 μg/h. Bottom: Baseline period

Lisuride was solubilized in distilled water and increasing concentrations of the solution were injected subcutaneously in the abdomen by means of the infusion pump, worn externally.

Three and a half milliliters of the solution were administered in pulses every 2.45 minuted over 24 hours.

The initial dose was 0.35 mg which was increased to 0.5 the 2nd day with further 0.5 mg increases every 3 days.

The site of the injection was changed every 4 days.

Results

As shown in Table 2 and Figs. 1–6, in all except case 2 we were able to maintain increasing doses of SCL over the first 5 weeks. Doses ranged from 41 to 104 µg/h with a mean of 69 during the fifth week.

Marked increase in mobility was observed throughout.

The mean "Off" disability score improved from 67.25 to 26.35 after treatment. In addition to the quality of "On" and "Off" periods, an increase in On period duration was also observed.

No major change in severity or duration of dyskinesias was evident except in case 4, in whom severe biphasic dyskinesias markedly improved with SCL (Fig. 6). As a noticeable improvement in mobility was observed during the first few days of treatment in case 1 (Fig. 4), the total levodopa dose was halved in order to decrease the severity of peak of dose dyskinesias, but a return was made to the previous dosage, due to worsening of parkinsonian symptoms.

Side effects included subcutaneous nodules 2–3 cm diameter developing at the injection site, in every patient, as early as 1–2 days following needle placement. However, local inflammatory reactions slowly disappeared over several days on changing the site.

In case 3 a subcutaneous hematoma 2.5 cm in diameter over a nodule was observed, and in case 4, several petechiae were evident. In both cases a complete hematological study failed to reveal clotting abnormalities.

Nausea was present in case 4 for the first few days of treatment.

During her 1st week of treatment case 2 accidentally received the whole daily dose of lisuride (0.5 mg) within a few hours, developing visual and auditory hallucinations accompanied by a paranoid delusion with marked aggressiveness. Symptoms cleared up slowly during the 4 days following lisuride withdrawal and levodopa reduction.

A week later levodopa dosage was restored to the previous level and bromocriptine added, but recurrence of psychotic symptoms led to her being excluded from the study.

Discussion

The present study was undertaken in order to determine whether continuous dopaminergic stimulation is capable of inducing a sustained homogeneous response in parkinsonian patients with otherwise major motor fluctuations.

Table 2. Results of subcutaneous lisuride infusions

| Patient | Time on treatment (weeks) | Drug therapy | | Off status (hours) | | | | Disability score | | Side effects |
		L-DOPA + DDI mg/day	Lisuride μg/h	Control	Infusion			On	Off	
1	5	500/50	104	10	2.5			23	42	—
2	0.5	687/68	41*	9.3	9			32	56	visual-auditory hallucinations, paranoid delusion
3	4	300/75	41	11	3			11	21	—
4	3	650/162	62	6.5	2			8	16	mild nausea

* 1 day

Mobility was improved in all cases as shown by increased quality and duration of On periods.

Furthermore, severe biphasic dyskinsias in patient 4 remitted considerably within the first week of treatment (Fig. 6).

However, both local and psychiatric side-effects may prove a serious threat. Subcutaneous nodules appearing throughout may hinder long-term treatment. Two patients (cases 3 and 4), also had hemorrhagic self-limited signs, though whether these were related to lisuride is uncertain.

Auditory and visual hallucinations occurred in one patient who also had a low Mini Mental Status Examination score, but an aggravating factor could certainly have been the rapid administration of the total daily lisuride dose.

Our results support those previously advanced by Obeso et al. (1986) and Castro Caldas et al. (1986) and confirm the usefulness of SCL infusions in the treatment of severely handicapped PD patients.

More cases treated on a long-term basis are required to draw definitive conclusions.

Acknowledgement

The authors are grateful to Mrs. Patricia Cabo for skillful assistance.

References

Barbeau A (1975) Long-term side effects of levodopa. Lancet 1: 395

Barbeau A (1972) Long-term appraisal of levodopa. Neurology (Minneapolis) 22: 22–24

Barbeau A (1974) The physiology of side-effects in long term L-dopa therapy. In: Mc Dowell F, Barbeau A (eds) Advances in neurology, vol 5. Raven Press, New York, pp 347–365

Calne DB, Claveria LE, Allen J (1974) Plasma levodopa and the "On-Off" effect. In: Mc Dowell F, Barbeau A (eds) Advances in neurology, vol 5. Raven Press, New York, pp 366–384

Castro-Caldas A, Costa Cristina, Sampaio Cristina (1986) Lisuride infusion pump for Parkinson's disease (Letter). Lancet 1: 1150–1151

Cotzias GC, van Woert HH, Schiffer LM (1967) Aromatic aminoacids and modification of parkinsonism. N Engl J Med 276: 374–379

Fahn S (1974) "On-Off" phenomenon with levodopa therapy in parkinsonism. Neurology 24: 431–441

Garcia de Yébenes J, Avila C, Bazán E, Gervás J, Maseda C, Mena MA, Muradas V, Ramos JA (1982) Aspectos farmacocineticos y farmacodinámicos del tratamiento con L-dopa, más inhibidor de la dopa decarboxilasa y agonistas dopaminérgicos en la enfermedad de Parkinson. Med Clin 78: 259–264

Gervás J, Muradas V, Bazan E et al (1983) Effects of 3-OM-dopa on monoamine metabolism in rat brain. Neurology 33: 278–282

Glough CG (1982) "On-Off" syndrome in Parksinon's disease and intravenous levodopa. Lancet 2: 765

Hardie RJ, Lees AJ, Stern GM (1984) "On-Off" fluctuations in Parkinson's disease. A clinical and neuropharmacological study. Brain 107: 487–506

Lhermitte F, Agid Y, Signoret JL, Studler J (1977) Les dyskinesies de debut et fin de dose provoqués par la L-dopa. Rev Neurol 133: 297–308

Marsden CD, Parkes JD (1976) "On-Off" effects in patients with Parkinson's disease on
    chronic levodopa therapy. Lancet 1: 292–296
Marsden CD (1977) The need for an alternative therapy in Parkinson's disease. In: Lakke
    JPWF, Karf J, Wesseling H (eds) Parkinson's disease. Concepts and prospects. Excerpta
    Medica, Amsterdam, pp 117–120
Marsden CD, Parkes JD, Quinn N (1982) Fluctuations of disability in Parkinson's disease.
    Clinical Aspects. In: Marsden CD, Fahn S (eds) Movements disorders. Butterworth,
    London, pp 92–122
Nutt JG, Woodward WR, Hammestad JP, Carter J, Anderson JL (1984) "On-Off" phe-
    nomenon in Parkinson's disease. Relation to levodopa absortion and transport. N Engl
    J Med 310: 483–488
Obeso JA, Luquin MR, Martinez-Lage JM (1983 a) Lisuride infusion for Parkinson's
    disease. Ann Neurol 14: 134
Obeso JA, Martinez-Lage JM, Luquin MR, Bolio N (1983 b) Intravenous lisuride infusions
    for Parkinson's disease. Ann Neurol 14: 252
Obeso JA, Luquin MR, Martinez-Lage JM (1986) Lisuride infusion pump: a device for
    the treatment of motor fluctuations in Parkinson's disease. Lancet 1: 467–470
Obeso JA, Luquin MR, Martinez-Lage JM (1986) Intravenous lisuride corrects oscillations
    of motor performance in Parkinson's disease. Ann Neurol 19: 31–35
Perret J, Ferrestein CL, Pellat J, Seore F, Gabend M, Touche M (1977) Resultats des
    dosages de la methoxy dopa plasmatique chez les parkinsonians induites par la L-dopa.
    Rev Neurol 133: 627–636
Presthus J, Holmsen P (1974) Appraisal of long-term levodopa treatment of parkinsonism
    with special reference to therapy limiting factors. Acta Neurol Scand 50: 774–790
Quinn N, Parkes JD, Marsden CD (1982) Complicated responses fluctuations in Parkinson's
    disease: response to intravenous infusion of L-dopa. Lancet 2: 992
Quinn N, Marsden CD, Parkes JD (1982) Complicated responses fluctuations in Parkinson's
    disease: response to continous intravenous infusion of L-dopa. Lancet 2: 412–415
Quinn N, Parkes JD, Marsden CD (1984) Control "On-Off" phenomenon by continous
    intravenous infusion of levodopa. Neurology 34: 1131–1136
Shoulson I, Glaubiger GA, Chase TN (1975) "On-Off" response clinical and biochemical
    correlations during oral and intravenous levodopa administration in parkinsonian pa-
    tients. Neurology 25: 1144–1148
Tolosa E, Martin WE, Coben JP, Jacobson RL (1975) Pattern of clinical response and
    plasma dopa levels in Parkinson's disease. Neurology 25: 177–183

Authors' address: Dr. Federico Micheli, Department of Neurology, Hospital de Clínicas
José de San Martín, University of Buenos Aires, Olleros 2240, 1426 Buenos Aires, Argentina.

J Neural Transm (1988) [Suppl] 27: 85–90

Lisuride infusion pump in Parkinson's disease. A report of two cases

O. S. Gershanik, O. Scipioni, and **Silvia García**

Division of Extrapyramidal Diseases, Department of Neurology, Hospital Francés,
Buenos Aires, Argentina

Summary. Two patients, ages 66 and 72, with complications of chronic levodopa therapy (random fluctuations, end of dose deterioration and dyskinesias) who were treated with Lisuride by means of a portable subcutaneous infusion pump are reported.

Results obtained show significant improvement in disability through a net increase in the number of hours spent "on". Dyskinesias remained unmodified. Limiting psychiatric side effects were observed in one of the patients.

Practical and technical aspects of the management of this therapeutic method are discussed.

Introduction

Among the methods used to achieve stable plasma levels of dopaminergic agents in the treatment of Parkinson's disease, the subcutaneous infusion of Lisuride by means of a portable pump as reported by Obeso and coworkers (1986) has attracted the attention of researchers worldwide.

It is our intention to contribute with our own limited experience in the use of this method in the treatment of parkinsonian patients with severe fluctuations in the response to orally administered antiparkinsonian drugs. We will be reviewing the results obtained in two patients who were given the soluble ergot derivative Lisuride by means of a portable subcutaneous pump over the last 4 months.

Patients and methods

Two patients, ages 66 and 72, with idiopathic Parkinson's disease, presenting with severe random fluctuations, complicated end of dose-deterioration, and abnormal involuntary movements were selected for the present study. Duration of symptoms were: 15 years for patient no. 1 and 9 years for patient no. 2. Both patients showed a good response to oral levodopa plus a decarboxylase inhibitor. However, in the case of patient no. 1, in addition to severe fluctuations, she experienced involuntary movements of the ballistic type as a benefit of dose dyskinesia. Thus, her "on" periods were as incapacitating as the time spent "off".

Table 1. Clinical characteristics of the patients

		Patient No. 1	Patient No. 2
Age		66 years old	72 years old
Years on levodopa therapy		14 years	9 years
Disability	"On"	29	14
scores	"Off"	57	62
Type of motor fluctuation		complicated end of dose deterioration "on/off"	end of dose deterioration "on/off"
Dyskinesias		ballistic biphasic	choreatic dystonic benefit of dose
Previous	L-DOPA/DDI	375 mg/day	875 mg/day
treatment	Other	lisuride (2 mg/day)	lisuride (0.6 mg/day)

"Off" periods were quite disabling in the two patients because of the severity of their parkinsonian symptomatology ("off" period scores were 57 and 62 respectively, King's College Hospital Disability Score).

The "off" periods were quite frequent and sometimes unpredictable, although the cumulative "on/off" charts (data not presented), showed a definite time relationship with each L-dopa dose.

Both patients were admitted to the hospital for a period of 10 days prior to starting the Lisuride infusions. They were evaluated on an hourly basis for several days, and instructed to fill "on/off" charts. After the initial assessment period, the patients were started on the pump. A battery operated Tecensa pump (model 62, Spain) was used, capable of delivering regular pulses every 3 minutes for 24 hours. The pump capacity allowed solutions up to 3.5 ml. The syringe was connected to a 16 gauge "butterfly" tube and needle. The needle was placed subcutaneously in the skin of the abdomen. Lisuride 0.5 mg ampoules were diluted in 3.5 normal saline and given over a 24-hour period at a constant rate of infusion. The initial dose of Lisuride was 0.5 mg/day, and was increased daily according to the benefit obtained. Both patients reached an acceptable level of performance with Lisuride doses of 1.5 mg/day. The daily dose of levodopa was gradually reduced as the dose of Lisuride was being pushed up.

Table 1 summarizes the clinical characteristics and treatment of the patients.

Results

Detailed results obtained with the infusions are given on Table 2 and correspond to week 3 of the infusion period. That was the last detailed evaluation made just before this meeting. Figures 1 and 2 are the "on/off" charts of patient no. 1 and 2, respectively. Data is given for the 10 days prior to starting the infusions and 10 days after reaching a stable rate of infusion.

Table 2. Results of subcutaneous lisuride infusions

		Patient No. 1	Patient No. 2
Time on treatment		3rd week	3rd week
L-DOPA + DDI (mg/day)		62.5 mg/day	625 mg/day
Lisuride (μg/h)		62.5 μg/h	62.5 μg/h
% time "off"	Control	76.2%	68%
	Infusion	3.9%	32.2%
Disability	"On"	11	4
scores	"Off"	22	36
Side effects		—	psychiatric** -confusion, -paranoid delusions

% time "off" is calculated on the basis of the waking period

Disability scores (King's College Hospital Disability Score) are calculated as the average of several assessments performed over a 10 day period

** Patient No. 2 was withdrawn from the study at week 5 because of psychiatric side effects

Both patients improved considerably. The reduction in the time spent "off" was dramatic having reached 95% for patient no. 1 and 53% for patient no. 2. There was also an improvement in the disability scores both in the "on" and "off" phases (Table 2). In the latter case, the patients were less severely disabled and tolerated better the "off" periods. Levodopa was reduced significantly in patient no. 1 (83%), while a moderate reduction (28%) was achieved in patient no. 2.

In regard motor side effects, patient no. 1 continued having severe ballistic dyskinesias (Grade 3 in a 0–4 scale), but they occurred only twice daily and for very brief period. This patient suffered severe and continuous dyskinesias whenever she attained therapeutic benefit prior to the infusion. In the case of the second patient, the benefit of dose dyskinesias (moderate in intensity, 1–2 in a 0–4 scale, and choreatic in type) persisted without modification. They were mild enough and did not functionally interfere with the patient's activities.

No other side effects were observed during the period reported here.

Patient no. 1 is still on the pump. Due to moderate deterioration in the therapeutic benefit, both levodopa and Lisuride had to be increased to 157.5 mg/day (divided in 8 doses of 31.5 mg/each) and 2 mg/day respectively. She has presently regained her previous status, now entering the 4th month on the pump.

Our second patient was withdrawn from the protocol on week 5. At that time, an increase in the dose of Lisuride to 2 mg/day, in an attempt to further smooth out the response, brought about a confusional state with paranoid

88   O. S. Gershanik et al.

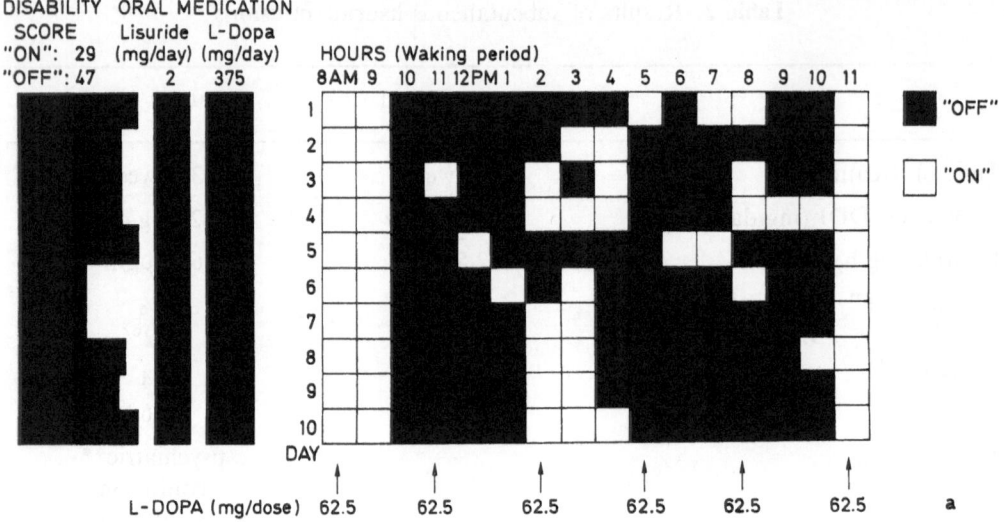

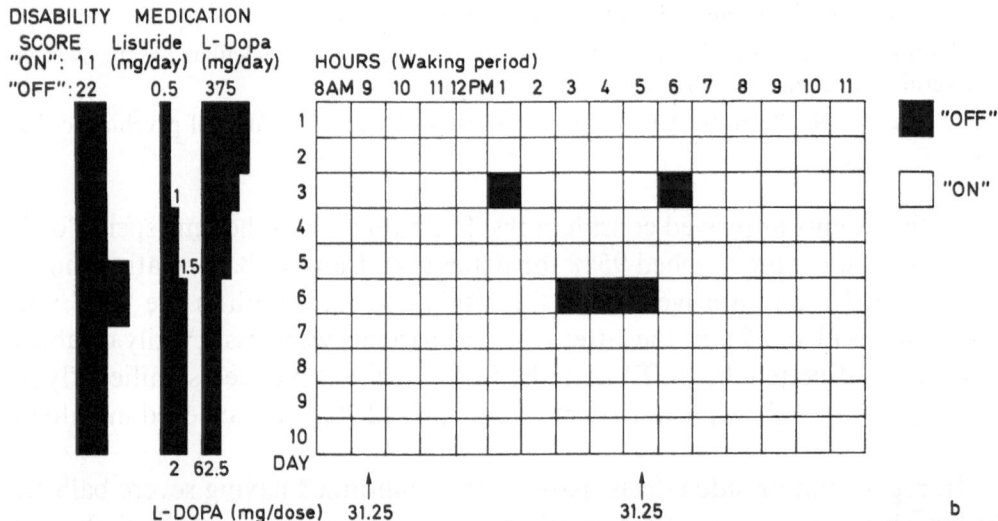

Fig. 1. Sequantial "on/off" charts of patients no. 1. Before (**a**) and after (**b**) Lisuride infusion. Disability scores (King's College Hospital Disability Score) are an average of several evaluations performed during a period of 10 days. Arrows indicate the time at which each dose of L-dopa/DDI was given

delusions. The patient's family did not accept to continue on the pump, even though it was explained to them that a reduction in the dose of Lisuride might result in the disappearance of the psychiatric side effects.

The occurrence of skin nodules in the injection site, which has been frequently reported in these patients, was a minor problem in our cases. Although these nodules, with overlying erythema and some irritation were present in our patients, they never required any special measures aside from changing the injection site every third day.

There were also some minor mechanical problems that are worth mentioning.

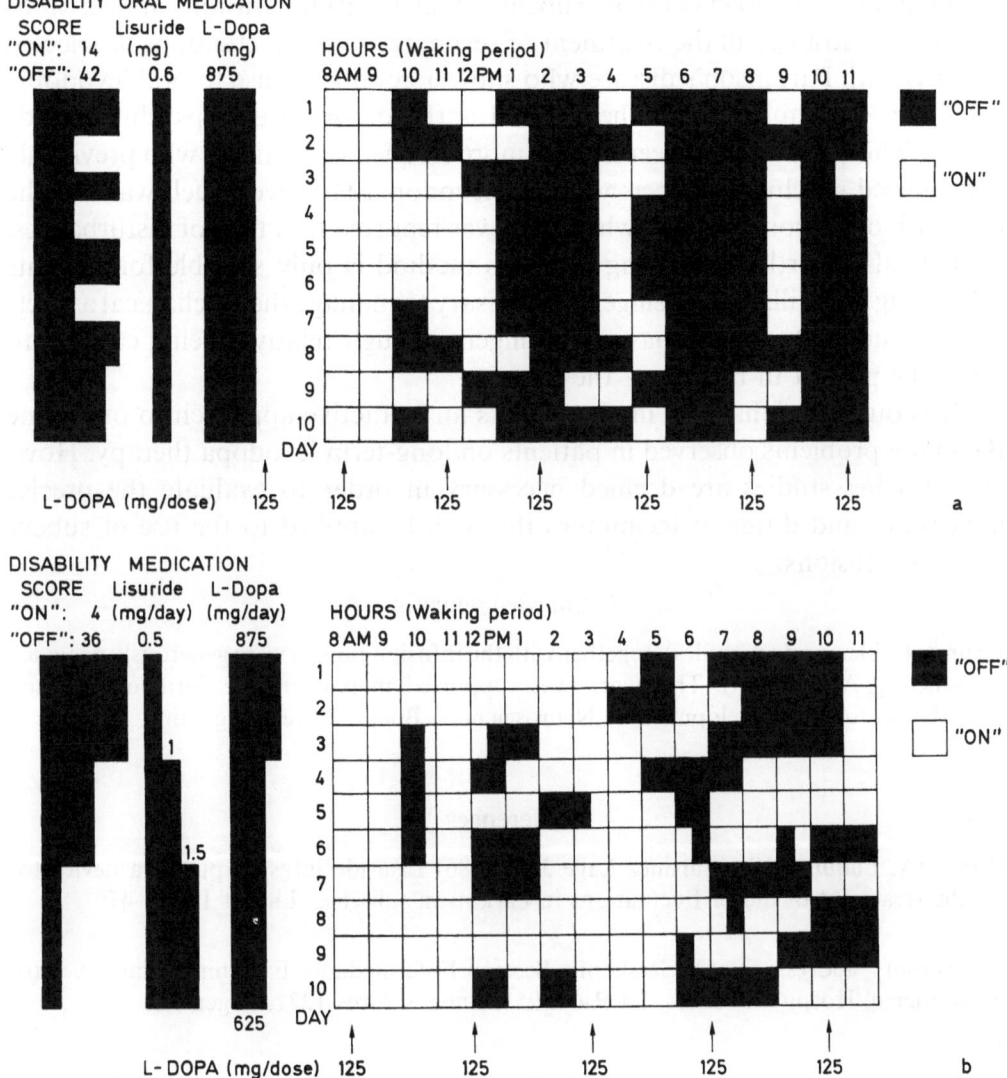

Fig. 2. Sequential "on/off" charts of patient no. 2. Before (**a**) and after (**b**) Lisuride infusion. Disability scores (King's College Disability Score) are an average of several evaluations performed during a period of 10 days. Arrows indicate the time at which each dose of L-dopa/DDI was given

In two instances, the pump was inadvertently turned off, with the subsequent worsening of the patients condition. In one of the patients, the pump driving device occasionally failed because of maladjustment of the mechanism. Both problems were easily overcome by careful observation from the part of the patients.

Discussion

The results obtained in our very limited number of patients confirm the observations made by several authors in this symposium, and the original results of Obeso and coworkers (1986). Subcutaneous Lisuride infusions appear to be an effective strategy in the treatment of severe response fluctuations in selected patients with Parkinson's disease who show a good response to oral levodopa.

The major limitation of the method is the occurrence of psychiatric side effects. This phenomenon seems to be more frequent in patients who previously experienced psychiatric effects while on levodopa. However, such was not the case with our second patient, who had never reported that type of disturbances.

It is also worth mentioning that this method is only suitable for patients possessing the skills and intelligence necessary to manage the mechanical aspects of the system. Or at least having intelligent enough relatives being capable to assist the patient in the use of the pump.

It is our belief that this method offers an attractive approach to overcome the many problems observed in patients on long-term levodopa therapy. However, further studies are deemed necessary in order to evaluate the precise indications and different techniques that can be applied to the use of subcutaneous infusions.

Acknowledgements

Lisuride (tablets and ampoules), together with the infusion pumps were generously provided by Schering AG (Berlin). This work was supported in part by the Alfredo Thomson Foundation for the Development of Neurosciences, Buenos Aires, Argentina.

References

Obeso JA, Luquin MR, Martinez Lage JM (1986) Lisuride infusion pump: a device for the treatment of motor fluctuations in Parkinson's disease. Lancet 1: 467–470

Authors' address: Dr. O. Gershanik, Sección Enfermedades Extrapiramidales, Centro Neurológico, Hospital Francés, La Rioja 951, Buenos Aires 1221, Argentina.

J Neural Transm (1988) [Suppl] 27: 91–114

Discussion

Management of motor fluctuations through maintenance of stable plasma levels

Fahn: I wonder whether the two speakers can tell us the composition of the dopa solution?

Nutt: We prepare the levodopa as follows: First we store it in ampoules containing 10 mg/ml in a solution that is made acid with hydrochloride generally at a pH of about 2. For administration we make it up in dextrose (5%), in water at 1 mg/ml. We bring the pH up to approximately 5.

Chase: We have a very similar preparation. I am not sure that we bring our pH as high as 5, I think it is more like 4, but nevertheless the preparation is very similar.

Let me point out that this is toxic to peripheral veins. The reason why we can continue for weeks is because we have to switch veins, and we run out of veins (and the good will of the patient) after a while.

Klawans: Have either of you any reasonable data on clinical response and/or levels of Sinement CR IV, which in our hands and others gives a better and more sustained clinical response associated with better blood levels?

Nutt: CR IV is not going to be a very long-acting drug, I think, but it is certainly more predictable and it is longer acting than regular Sinemet.

Gessa: Do you know whether carbidopa crosses the blood-brain barrier to some extent and therefore causes some impairment or changes in the formation of dopamine?

Chase: We have looked at this in the early seventies, when carbidopa was called MK 485 and MK 486. We increased the dose something like 10 times above that currently used, and could see no inhibition to suggest that this is really getting into the central nervous system in any significant quantities. So, I think, there is a wide margin of safety for carbidopa.

Gessa: This could be checked by measuring, for instance, the dopamine metabolites in the cerebrospinal fluid.

Poewe: I'd like to ask Dr. Nutt about 3-0-methyldopa. From your talk I understand you don't really think that the accumulation of 3-O-methyldopa during long-term infusion is of any major singificance with respect to interference with clinical effects. On the other hand, amino acid or protein loads on the second day of infusions led to greater clinical deterioration. Do you think that the higher 3-O-methyldopa levels that you probably had during your second day might have anything to do with the greater sensitivity to amino acid loads during the second day?

Nutt: Let me clarify that. 3-O-methyldopa can clearly inhibit the action of levodopa, and we have shown that during long infusions. During the short infusion we gave an oral challenge of 3-O-methyldopa and reduced the clinical response. However, to do that we were giving very large doses of 3-O-methyldopa, 100 mg/kg, and were producing plasma levels that averaged 600 nmol/ml. The normal concentration of 3-O-methyldopa during oral dosing is in the vicinity of 15–70 or 80 nmol/ml. So, we were producing concentrations that were 10 to 20 times higher than we have ever seen in any of our patients on regular oral therapy. 3-O-methyldopa is only one of many plasma amino acids that is competing for transport. The other amino acids have a total concentration in the vicinity of 500–700 nmol/ml. So, going from 20 to 40 nmol/ml of 3-O-methyldopa is really a small contribution to the total amount of large neutral amino acids that are competing at the blood-brain barrier. That is why we think quantitatively the 3-O-methyldopa probably under normal circumstances is not a critical determinant. The other thing that we looked at was whether 3-O-methyldopa did have a higher affinity for the transport system than the other amino acids we compared it in a crude way in our patients with phenylalanine, but it was not more potent.

Stahl: Dr. Chase, did you think that the 14 day infusions ever caused something you'd call tolerance? And particularly when you saw that the fall off in the reduction in variance in the on-off patients which wasn't as robust as the wearing-off, did you think you were at the end of the party, or that if it would have gone two months you would continue to get more reduction in variance?

Chase: In retrospect I think that we should have carried out the experiment longer. If you look at the curve, it appears asymptotic down to near zero, but it really hadn't achieved that. I think that perhaps if we had continued longer we would have gotten further reductions in variance. In terms of tolerance, we don't really see tolerance to dopa over the periods of infusions that we have given.

Melamed: I wonder if you could care to speculate a little bit more about this phenomenon that you showed us of suppression of motor activity by giving lower doses of levodopa. You suggested that there may be some dual effect of levodopa during low doses and higher doses. This is possible. Could you consider that, at lower doses levodopa turns down the firing rates and turnover of dopamine from the surviving dopaminergic neurons, which somehow collaborate with the dopamine that is formed in other compartments in the striatum?

Nutt: You are probably better equipped to make those type of speculations than I am, but I guess the concern that I would have is that there isn't that much of presynaptic mechanisms left in these patients. So, I would favour moving the mechanism to postsynaptic, perhaps something to do with D1/D2 receptors. Another suggestion that came from a pharmacodynamicist, is that here may be two different compartments, and different rates of entry and exit from those compartments could give you this type of biphasic response, one compartment perhaps being putamen, the other being caudate. So, you could put together mechanisms like that as well. I don't know, but I am rather fascinated by this phenomenon because I think a lot of the patients' dose to avoid the "off". Some of those patients will tell you that if they can wait it out that they will be much better 2 or 3 hours later without taking any medication. I think that if we had a therapeutic way of abolishing that down-period, that acute off-period, that the patients would do much better.

Chase: We have not done the experiment exactly the same way Jay Nutt has. We have tried dose response testing using single i.v. boluses of increasing amounts. What we see is that we can increase the dose and get increasing response. We have not seen an inhibition of response at relatively low doses. But the way we do the experiment wouldn't lend itself to seeing that kind of change. What we do see, however, is that the therapeutic window narrows in our patients. And I think this is really critical. Early on in mild patients you see a big range between the minimal dose which obliterates Parkinson signs and the dose that is required to produce dyskinesias. In the more severe patients this window narrows down to zero, in fact inverts, and you can get dyskinesias at i.v. doses which do not yet completely eliminate Parkinson symptoms.

Horowski: I am very interested in this rebound of rigidity or parkinsonism you have observed. There is an analogy. If you infuse dopamine to hyperprolactinaemic patients you will lower the prolactin level, and by stopping the infusion you get a very severe rebound of prolactin, high above the pre-existing plasma levels. You won't see a similar thing when infusing dopamine agonists of the ergot type, such as lisuride. So I am wondering what will happen when you stop the lisuride infusion. One could speculate on several possibilities: An antagonistic system that remains active for a longer period of time as a kind of tolerance. Another possibility in the case of levodopa would be a dopamine metabolite with a longer half-life, such as 3-0-methyldopa, which has some antagonistic properties. Of course, also autoreceptor stimulation must be considered a strong probability.

Chase: Another possibility is that lisuride has some antagonistic properties and may be eliminating the postsynaptic blockade, and allowing more the endogenous dopamine to interact with the postsynaptic receptors.

Nutt: I don't think it is going to be unique to levodopa because we have seen this phenomenon in patients that have received challenges of apomorphine. Maybe it takes D1/D2 receptor agonism to produce that. It is not going to be just a levodopa phenomenon.

Obeso: I want to add that it also may have to do with the way the drug is given, whether it is acute, chronic or subacute.

Quinn: I think you'd necessarily have to have the "on"-period, because I think we have all witnessed the fact of the patients' first sign that they are beginning to turn on is that their tremor gets a lot worse. Secondly, patients who were initially treated with low doses of dopa before they start to get the benefit may come back saying I am not taking any more the stuff, it's making my tremor worse. I am sure, we have all seen that.

Nutt: I agree with you on your observation about the tremor, except I was reluctant to make that type of comment based on tremor alone. I really wonder if tremor and dyskinesia are very closely related. A lot of patients as they begin to turn on develop more severe tremor and then will switch right into dyskinesia. I would be more impressed when we could show that phenomenon for other parameters of parkinsonism, other than simply the tremor.

Marsden: There is also another confounding factor about tremor and that is the potential for peripheral beta-stimulation that might make tremor worse before the central dopaminergic stimulation starts to make it better. I don't think tremor alone can be used as a pure index of central dopaminergic effects.

Hardie: Just to make two or three observations: Firstly, the question of this apparent rebound or deterioration in effect. This is very reminiscent of my own studies. The protocol we followed was to titrate the dose response and this entails starting with a low dose of levodopa given by a constant intravenous infusion. After slowly increasing it, I would come across a level where the patients actually switched on, and if one was patient they would stay on for a few hours and then switch off. I kept thinking they are bound to turn back on and one could maintain that level for the rest of the day and literally there will be no further response. I think it is not absolute levels which determine the kinetics of any carrier mechanism. The key thing is the rate of change, and I think that this is something that we have to take into account whenever we are trying to correlate absolute plasma levodopa level at time x with clinical motor response at time x. As we know, you cannot interpret it without knowing what the plasma level was at $x - 10$ and $x + 10$. Rate of change of plasma levodopa level must be just as important as absolute levels.

Nutt: Actually I don't know what to say about rate of change. Certainly it is not absolute levels but ratios of plasma levodopa to other competing amino acids. I don't know enough about the kinetics of rate of change to know how that might influence things. But I think the other point you are bringing up is very important, that there is some sort of lag in the system. You may not find an instantaneous correlation between the plasma levodopa levels or the plasma levodopa/amino acid ratio and clinical response. You really need to know what has been happening over the preceding half hour or hour. The response on the rising level of the plasma levodopa curve is quite different from that on the falling level of the plasma levodopa curve.

Hardie: Can I make a couple of more points to Dr. Chase. One point that you made that intrigued me was that the levodopa requirements appeared to be stable irrespective of time of day or night, and mobility versus repose. That would not necessarily tie in with my own impressions. I just wonder whether it is because your plasma levodopa levels are stable at rather higher levels than those which Jay Nutt used and are much higher than those which I used. Perhaps if one is close to threshold then mobility and time of day will make all the difference, whereas if you are way above it then perhaps that might explain it.

Chase: What I meant to imply is that we could set the pump at an optimal rate and maintain that rate day and night. It didn't seem as though we needed to continue readjust the rate at which dopa was being infused in these ambulatory patients. We do see what Jay Nutt was alluding to before, and that is that some of our patients in the afternoon don't do as well as in the morning. And we wondered about what that problem is, whether they were sleepy or whether it was lunch or what the factor was. Ordinarily, with the diets that our patients get in hospital we see no effect of the diets. However, our classic patient was a patient who went home on a pump and had that holiday dinner with his family and immediately turned off. So, obviously you can eat enough to change the response to a constant i.v. infusion of dopa, but ordinary meals don't seem to do this. On the other hand, if we measure large neutral amino acids over a 24-hour cycle we find that there is a circadian rhythm and that these go up in the afternoon. And we are just wondering now whether this rise in large neutral amino acids in the afternoon could in some way impede the uptake of dopa in the CNS and cause this afternoon reduction in response that we see in some of our patients. I don't know whether anybody else has had the same experience.

Hardie: Can I just come back on that? One of the things that we know that alters the uptake of amino acids across blood-brain barrier is insulin. And obviously insulin levels are changing all the time. It may well be that that is one confounding factor and there are others.

Melamed: We were also fascinated by the fact that levodopa seems to be working less well in the afternoons and evenings than in the mornings in the majority of patients. Occasionally a patient would say that he enjoys dopa better in the afternoon, but that is a rarity. So, we monitored plasma levels and clinical responses, both after the morning dose of levodopa and after an evening dose of levodopa. We found actually no differences in the various parameters that we monitored. This suggested that differences in the response to levodopa in the mornings and the afternoons are probably not due to peripheral mechanisms, but something happens in the brain itself to the utilization of dopa or perhaps to postsynaptic mechanisms. So it fits with your idea.

Duvoisin: The two gentlemen have the equipment to infuse other compartments than the cardiovascular. Have you, for example, tried the intestinal tract along the lines of Kurlan and Shoulson and Jacob Sage and my department?

Chase: We have not done that.

Nutt: We are doing that in conjunction with Roger Kurlan and we have infused both intraduodenally and intragastrically. We are impressed that we get very stable plasma levels with infusions in both compartments, which surprised us. We expected it to be quite erratic in the stomach and constant intraduodenally.

Quinn: You just reminded me of this problem of rate of change and absolute levels. In the infusion paper we did in "Neurology" we gave a striking illustration of a patient with diphasic dyskinesias. Every time you gave a big bolus, an acute increase in levodopa, you got a flurry of dyskinesias and then the patient turned on. Then after being on for a period of one and a half hours or so, as the levodopa level was slowly falling, they go back into the end-of-dose dyskinesias. You could abolish those with a further bolus of dopa and you ended up going higher and higher and higher. This patient got extremely toxic on the $18 \mu g/ml$ he ended up at. Maybe it's something to do with ratios and that is ratios with 3-O-methyldopa rather than absolute levels per se.

Tolosa: Many of my patients feel better in the afternoon than in the morning. I think, in Spain it is possible that Siesta plays a role, some recuperation, and that may be a consideration. The other thing I would like to underline is what Dr. Hardie has said about the rate of change. When we initially studied plasma dopa levels and fluctuations in motor performance, we found that patients with very high levels of plasma dopa had very severe off-periods when they had rapid changes from high levels to low levels. Here I would agree fully with Dr. Hardie that the rate of change must play an important role.

Duvoisin: Jacob Sage and my group have been introducing levodopa solution via a nasoduodenal tube, or directly through a feeding gastrostomy tube, and find that we get results that are remarkably rapid. The effect is almost as rapid as intravenous administration. In the dog models with which we worked, it is very striking that the first time we placed a bolus of levodopa solution into the jejunum, the dog vomited within a minute or so. That is as rapidly as would have occurred with an intravenous injection. We have 6 patients now, and the duration of treatment is from two weeks at this moment to a little over two months. In the latter case of a man with a feeding gastrostomy it has been very effective. He walks around with the same plastic jacket that you have shown and with a similar pump mechanism. It has converted a totally disabled individual into a functional person who can go to the theatre and go out to eat in the restaurant and function in a virtually normal way receiving this infusion around the clock in the same manner as you described with the intravenous one. It seems to circumvent many of the problems with intravenous use.

Fahn: Just to get back to Jay Nutt's presentation, I am troubled about the point where you state that only the duration of response is related to the dose and not the magnitude of the response. And I just wonder if it is because the response that you measure like walking has already reached maximum and you can't get any further improvement in that particular patient or other patients. Because surely when you look at toxicity, let us say dyskinesia is toxicity, you can increase the dyskinesias by giving higher and higher doses each time. And that seems to be some type of magnitude of response related to dose, not just to duration only.

Nutt: All I can say is that what we have generally seen has been again that magnitude seems to be relatively independent. And even when the patient comes on for at least a short time with a smaller infusion rate, the dyskinesia often looks very similar to what would be seen if the response lasts for several hours. My impression is that dyskinesia may increase with the duration of infusion, but often the plasma levels aren't increasing. So there may be something that happens with dyskinesia if it goes on for longer periods of time. But to my surprise, we did not see any clear-cut increase in dyskinesia with doubling or tripling our threshold infusion rate. Occasionally in patients you can see that, say, the tapping scores incremented if you go up a little bit higher, but generally we don't. It really is duration of response that seems to be most responsive.

Przuntek: Dr. Obeso, were you able to reduce the dosage of lisuride during the infusion time? For example, if you started with 2 mg per day, were you able to reduce the dosage during one year?

Obeso: No, if I have done that it is due to side effects, not due to enhanced sensitivity.

Nutt: A comment and a question. First, the variable plasma levels may represent variable absorption and, at least as far as I understand, absorption from subcutaneous sites is primarily determined by blood flow. The question is, have you tried any patients with lisuride alone? Can you get away from using levodopa in these patients?

Obeso: Yes, we tried hard. Two patients were very different from the whole group because they had juvenile Parkinson's disease. These are the typical patients who are very difficult to get really well controlled, either they are dyskinetic or they are off, even with minute doses. The two of them at a certain period during the treatment had been without levodopa. In one of the patients it was for a few months when she developed psychiatric complications, and the other patient kept going for a little while and then we realized he needs a little bit of levodopa. He is now taking about 62.5 mg at night, but he needs something, otherwise he doesn't feel really good.

Chase: I think it is a very common experience that none of the ergots are quite as good as levodopa. That is the subjective feeling that our patients report.

And I can't put my finger on exactly what the problem is. Why aren't they quite as good? Is this because of the D 1/D 2 ratio or is this some other factor?

Tolosa: I want to comment. The slide José Obeso showed with lisuride plasma levels following oral administration, seemed to say that the clinical response was not as good as with similar plasma levels with subcutaneous lisuride. Is that correct? The question is was this patient also taking levodopa or not? Because it seems to me that the response of this agonist is different, even though you may reach the same plasma level, if the patients are taking levodopa as well.

Obeso: They were taking levodopa, the same dose, in either condition. The problem is that, except in the slide I showed, the plasma levels after oral administration are much lower. The total oral amount of lisuride you need to give to the patients to reach the same plasma levels seems to be so high that we just were a little bit afraid of doing that.

Tolosa: It seems that the route of administration from your studies is important, whether it is oral or i.v. or subcutaneous. Then the concomitant administration of levodopa is another factor that seems to be important for the effectiveness of lisuride. I just would like to comment that the same is true for an agonist like apomorphine. So, there is nothing peculiar to this ergot derivative. At the time when I was working with George Cotzias, we had a similar experience.

Melamed: The patients are monitored in the hospital, not in the normal situation. I wonder whether you, José, and Tom also, who gave these infusions of levodopa on a long-term basis could comment on the requirement of the drugs as related to the physical activity of excitement. For instance, I have a group of patients who are physical labourers. It is quite interesting that when they are working physically hard labour, they consume, they tell me, the levodopa faster and the effect lasts shorter than when they are relaxed or resting. And I wonder whether this is also your experience.

Obeso: Regarding lisuride, our policy so far has been to try to keep constant infusion rates. Therefore we have cautioned the patients not to change the infusion. Some of them do realize that with a little bit more they could be better, but there hasn't been, as far as I am aware, a clear-cut correlation. For instance, that's why I wanted to show the man skiing. He doesn't increase the infusion rate to go skiing, as far as I know.

Chase: We have seen no increased requirement of levodopa with exercise in a standard exercise tolerance test.

Quinn: We have just recently been through all of our young onset parkinsonian patients below the age of 40, we have 56 between the age of 25 and 40. One of the extraordinary things about them is that after one year of treatment 50% have got dyskinesias and fluctuations, and after 5 years 100% have got these problems. So they are particularly at risk for developing all of these complications. And if we are going to show any difference in how we can avoid these problems by using different treatment regimens, I think this is really the group to concentrate on.

Stocchi: It should be interesting to look at these patients because even our patients, though not as young as yours, develop a levodopa long-term syndrome, with on-off fluctuations and dyskinesias, much earlier than the older. So I think, giving a continuous infusion to these patients early in their disease could be very interesting.

Rinne: You suggested giving this kind of infusion to de novo patients who have never received levodopa. According to my long-term experience, giving a dopamine agonist alone is very good as far as the development of fluctuations is concerned. On the other hand, the therapeutic response is not so good. What is your reason to try this kind of infusion in de novo patients; why not to treat them orally?

Stocchi: Talking about lisuride, our group has quite a long experience with oral lisuride treatment. I can tell you that the clinical response to oral treatment in comparison to a subcutaneous one is absolutely different. Patients who don't respond at all to lisuride orally can respond beautifully to the subcutaneous route. I think this is a good point to look at. The other thing is, that the infusion provides quite constant plasma levels, not so constant as we believed in the beginning, but much more constant than with oral treatment.

Rinne: Is the difference in therapeutic response between oral and infusion just due to the higher dose, or is there some other reason why infusions give a better therapeutic response?

Stocchi: I think you are right. With the subcutaneous infusion you can have a higher level much more easily than with the oral treatment. It is very easy to produce a constant flat plasma level by giving lisuride subcutaneously.

Marsden: The behavioural differences in animals from continuous bombardment of dopamine receptors as against pulsed bombardment are quite dramatic. One of the reasons for thinking about administering constant dopamine stimulation at an early stage of Parkinson's disease is that it may stop the development of some of the fluctuations, which may be due to these curious and unexplained pharmacodynamic differences that one can see with different modes of administration.

Chase: I would like to support that possibility. All our data would suggest that dopa plays a role in the onset of these motor fluctuations, and it may not be the dopa but the way it is administered, and that is in a pulsatory way.

Stahl: Your first conclusion was that constant drug levels do produce a constant effect. Is that really what you meant? Did you prove that constant drug levels were present, or is this a constant input rate?

Critchley: The constant level I was referring to was actually from the levodopa work and we don't have pharmacokinetic data from our own lisuride infusions. You are quite right, we were talking about a constant level of infusion as regards lisuride.

Przuntek: When we need L-dopa, why don't we infuse lisuride and L-dopa together? Then we would have, I believe, optimal levels of necessary substances.

Experience with subcutaneous lisuride infusion

Marsden: The first question I think we ought trying to attend to is which is the best drug to use if we are going to produce continuous stimulation of the brain. It is not yet at all clear to me from what we have heard this morning, whether lisuride is better, worse or the same as levodopa as a drug for use for continuous stimulation. I'll be interested to hear people's views. That is the first question, is lisuride better, worse or the same as continuous levodopa?

Yébenes: I don't know if I can answer that question because it depends on the concentration. Dopamine is the most potent drug available for stimulation of dopamine receptors in the brain and for obtaining pharmacological changes. But lisuride has a very peculiar effect. During the first days of infusion it is not so effective, and then the effectiveness increases after a few days, something like 4–5 days after initiation of infusion. The opposite has been true with levodopa.

Marsden: But Tom Chase clearly showed in his data, that there was an improvement over the first few days of his continuous levodopa infusions.

Chase: There seems to be some difference in efficacy between lisuride and other dopamine agonists and levodopa. The question is to find out exactly what that difference is. I think we have all had reports from our patients (in our case it is with bromocriptine, but I think that with other dopamine agonists the story is about the same) that using these dopamine agonists as monotherapy does not produce the same kind of antiparkinsonian response as levodopa itself does.

Gershanik: We have been working for many years with the dopamine receptor interactions, and in recent research work we have done we keep proving that you need a baseline D 1 stimulation in order to obtain a more physiological motor response. So I think the answer might lie there, that you need a very specific profile in an efficacious antiparkinsonian drug, i.e. that you have to stimulate both receptors in order to obtain the best response.

Marsden: Has anyone here got experience of comparing in the same patients continuous infusions of levodopa versus continuous infusions of lisuride?

Stocchi: Yes, we performed a continuous levodopa infusion in 20 patients, and in the same patients we gave a continuous intravenous infusion of lisuride.

What we found is that all the 20 patients responded brilliantly and satisfactorily to levodopa, but just 60% of these patients responded to lisuride. Another point is what the patients report about their on-state. With lisuride alone they are on and mobile, but they keep saying that their on-state is less brilliant than with levodopa. From the clinical point of view there is no difference in Columbia rating scale.

Chase: Since the question of relative ratios of D 1/D 2 potency has come up several times, I would like to report of some experiments that we have done with SK&F 38393. This is a very selective D 1 agonist. If D 1 stimulation is required in our Parkinson patients, then you think this drug given in adequate doses would produce some antiparkinsonian effect. We have found that giving it alone to our Parkinson patients at levels which produce elevations of the drug in spinal fluid, comparable to what we find in the brain of a rat when the rat shows some motor responses, there is no antiparkinsonian effect. And moreover, when we give i.v. dopa at either sub-optimal levels or slightly super-optimal levels we cannot modify the response to i.v. dopa by adding on SK&F 38393. I think that although I would like to believe that you have to have drugs which are equipotent on the D 1 and D 2 systems, as after all dopamine in fact is, we can't prove that with the pharmacological tools which are currently available.

Gershanik: What about the infusion of lisuride plus SK&F 38393?

Marsden: Niall Quinn was about to ask the same question. Dr. Chase says he has no data yet.

Leenders: Is it not simply that for levodopa you have an additional compartment, particularly in the early stage of Parkinson's disease, i.e. you have an uptake and a storage, whereas with the other agonists, there is no such a thing. Do you know whether in far advanced cases, where apparently the nerve terminal pool will be much less than in the beginning, it is the experience that levodopa still has a much better response than agonists, or does that apply only to early cases?

Marsden: I return the question to Dr. Stocchi. Were your cases in whom you compared levodopa infusions with lisuride infusions advanced fluctuating cases?

Stocchi: Yes. The difference, even in responding patients, between levodopa and lisuride is that with lisuride there were still some fluctuations during the day. But there is another point I would like to make. In our patients treated for 12 hours with subcutaneous infusions with lisuride, we found out that the blood level of lisuride is almost stable, as José Obeso showed this morning. But a very interesting point came out from the blood samples taken in the following morning, 12 hours after discontinuing the infusion. We found that 0.13–0.15 ng of lisuride were still present in the blood. If we look at the half-life of lisuride we see that is much shorter. This is a point that I make to the pharmacologist who is involved. Maybe this aspect can explain why our patients even receiving 12 hours infusion during the day have quite a good night. The other point is, why are there some patients, quite severe, that respond satisfactorily to lisuride alone which is claimed to be just a D2 receptor agonist. There are other patients that don't respond at all or respond to lisuride infusion only if you give it combined with oral levodopa.

Obeso: My feeling is that the severity of the disease is an essential point. If you treat patients who are severe because they already are fluctuating, but not that severe, you can get away without levodopa. In fact, because nobody is answering your initial question, I would like to comment upon it. Perhaps one may not try to decide whether the question is levodopa versus lisuride. Perhaps the problem is which patient needs what and what is more practical in a given situation. The Rome study seems to indicate that lisuride may be practical and useful in a certain type of patient while levodopa also has a number of further complications. I would like to get a clear-cut statement form Jay Nutt and Tom Chase about what is the real significance of your data on the possibility of interference with levodopa absorption. Whenever I talk or write about that I always say levodopa is not going to be used in practice, because it can have such problems. However, you showed that it can go on working beautifully. Is that really a practical drawback of levodopa infusions, or can we forget about that and think levodopa has as good bioavailability as lisuride?

Marsden: Before we go further in the discussion, one has to remember the data Roger Duvoisin brought forward, i.e. the intragastric infusions of levodopa which get over all the problems of venous access and difficulties like that. It may have its own problems but it's another technique of delivery of levodopa.

Gershanik: Even though the differences in the patient population from Rome were not statistically significant, those patients that could be treated with lisuride alone were younger, less severe, with shorter duration of disease. You can speculate that they have more residual dopamine in their striata that could be responsible for the D1 stimulation that you need.

Gessa: There is evidence from experimental data that the effect of some ergots is conditioned by the presence of dopamine. For instance alpha-methyltyrosine plus reserpine eliminates the effect of bromocriptine. I don't know about lisuride, I don't have the same evidence. So there is a possibility that these ergots are partially working via endogenous dopamine. One thing that we found, and Dr. Mereu will present the data tomorrow, the major action of lisuride is to stabilize dopamine autoreceptors. Stabilize means that these receptors cannot be longer activated by apomorphine. One possibility is that lisuride plus dopa works because lisuride is preventing L-dopa from inhibiting the remaining neurons. So there is a possibility also that in younger and less severely impaired patients lisuride works, just as you said, because it is probably permitting endogenous dopamine to act postsynaptically.

Horowski: We did studies, as did many others, showing that bromocriptine no longer has effects on the motor system when there is no dopamine left. That is different from the prolactin system. Lisuride was still effective following alpha-methyltyrosine plus reserpine, but its efficacy was slightly reduced. I feel there are different classes of compounds, levodopa being one with a presynaptic uptake characteristic which may be very useful, but also with its interference in its uptake with other amino acids and most probably 3-O-methyldopa. Another one is apomorphine which is still the classical pure postsynaptic agonist. In the early seventies we did studies with Dr. Strian from Munich with infusions of apomorphine in parkinsonian patients. Those patients recovered very well; they got good improvement mostly on the tremor but the therapeutic range was too narrow, they quite frequently developed severe hypotension and other side effects such as emesis in spite of peripheral antiemetics and even psychiatric complications so this couldn't be followed up. That is different with the ergots. And what Prof. Gessa just has mentioned, the stabilizing effect that most of these ergots have might be different. We indeed find this with terguride much more pronounced than with lisuride. And in clinical practice now with oral treatment, I have the impression that combining levodopa with an ergot dopamine agonist not only improves motor performance, but also reduces hyperkinesias due to DA receptor overstimulation. So this smoothening effect might play a role, and so I would even think that, if possible, one should try to combine infusions of levodopa with lisuride.

Rinne: I agree with Dr. Horowski that combining low-dose levodopa together with a dopamine agonist seems to be a better treatment for parkinsonian patients than high-dose levodopa alone.

Marsden: Has anybody got experience of combining lisuride and levodopa infusions in the same patients at the same time?

Stocchi: We performed a continuous intravenous infusion of standard levodopa with subcutaneous lisuride in what we call nonresponder patients, and the result was beautiful. And even the control of dyskinesias was possible. That sometimes is not possible giving oral levodopa. From the motility point of view, the results were quite similar to those with oral levodopa combined with subcutaneous lisuride. But we have done another thing that is very interesting. We gave levodopa methylester subcutaneously together with lisuride subcutaneously, and we had very good results indeed from the motility point of view and from the abnormal movement point of view. The only problem was that so far, even if our methylester is very concentrated, 500 mg in 2 ml, it is not possible to give it subcutaneously because the patients develop after 12 hours red nodules comparable with those observed with lisuride, but more stable and more painful. This is the shortcoming so far, but I think we could look further at this idea. I would believe that the patients who respond to lisuride alone are less affected or that there is some dopamine still in their striatum. The problem is that we observed the same results in the patients that are really very severe, with un-predictable on-off fluctuation and terrible off-state; they respond beautifully to lisuride alone. Sometimes patients that are less affected, not very severe, don't respond at all. That is another point for the discussion. The synapse is a rather complicated structure. There are, for example, the cotransmitter- and the comodulator-peptides that play a very important role. I would like to hear some comment about this point, because in all our studies we forget about peptides and we just act on pre- and postsynaptic dopamine receptors.

Marsden: I think we also have to take into account in all these discussions the dose one is giving. Can someone tell me what the equivalent plasma concentration of lisuride during constant subcutaneous infusion would be to the concentration of levodopa in plasma to produce the same motor effect, in absolute terms?

Obeso: We have never done the intravenous levodopa infusion, but I would doubt that out of 100 patients doing the two infusions you will get 50% in whom you can compare, because the feeling is that you are never going to get the same movement capacity with lisuride alone. I don't know if you can do the comparison altogether. It could perhaps be interesting to ask for the experience of all the people who have been working with lisuride infusions to see how many patients altogether do we know who have been doing well without levodopa.

Marsden: How many patients have been managed on subcutaneous lisuride alone without levodopa? I calculate that the total number of patients at present were something like 60. So it's about 6 out of 60 roughly who had been managed on lisuride by itself, all the others have been with levodopa.

Chase: The next question that is not clear to me is if lisuride plus levodopa is better than lisuride alone, what is the proper ratio of the two drugs. Does anybody know? How much do you give at each in order to get the optimal response? There is a big variation in the dose ratios as we have heard this morning.

Horowski: You will have to answer this question yourself by doing clinical studies. I just want to make a statement on effective plasma levels. We feel that in those patients where you can achieve best motor reactions, plasma lisuride levels should approximate 1–2 ng/ml, whereas with levodopa the levels should be about 5000 times higher. So this would somehow reflect the receptor affinity.

Marsden: Can we just pursue that aspect of the conversation and press Fabrizio Stocchi to tell us more about that unique set of patients who are being infused with levodopa and at other times with lisuride. What were the comparative blood levels, remembering that the ratio has to be something like 5000 to get equivalent effects.

Stocchi: I can't speak about blood plasma level, but I can tell you that the lisuride infusion regimen per hour for the responder is 64.1 ng/hour and for the nonresponder (levodopa plus lisuride) is 81.4 ng/hour. I haven't got the blood levels so far.

Marsden: What was the infusion rate for the subcutaneous lisuride in those patients who also had intravenous levodopa; what were the comparative infusion rates?

Stocchi: I haven't got the data under my eyes but from memory that for levodopa in hours it is 55 μg/hour and for lisuride it is 64.1 ng/hour.

Melamed: How much carbidopa was the patient taking?

Stocchi: 25 mg every 4 hours.

Quinn: Could I ask these people who have been using lisuride alone whether they have been able to observe the same kind of threshold effect that we have observed with on-off fluctuations with levodopa. If you see such a threshold effect, do you think that that indicates that this is occurring at a postsynaptic level, or could you think you could explain it on doing something to destabilize or stabilize autoreceptors?

Stocchi: It is a very complicated question. We found out that the threshold in the three patients that we have got the lisuride blood samples is around roughly 0.5 ng/ml. This is the threshold to turn the patients on, but then to keep the patient on, the threshold is under 0.45 ng/ml roughly. I can't answer the second question because it is very complicated.

Horowski: May I just tell you that there is an important paper by Parkes et al. on intravenous lisuride at different dosages and concentrations, and so far I don't remember that there was this kind of inhibitory effect reported there. For me it was a basis because it was done intravenously at this time and it was very nice clinical pharmacology. It is a basis for what we are doing now with subcutaneous application.

Palacios: I am afraid you have to introduce another complicating factor in your calculations when you are talking about dopamine agonists. The activity of an agonist is not only related to affinity, it is also related to intrinsic activity or efficacy. I think that lisuride is a partial agonist, and the activity of a partial agonist, dependent on the status of affinity or sensitivity of the receptors, will be different from area to area, from patient to patient. This is an important factor that has to be kept in mind when you are doing all this type of calculations.

Horowski: I can't but agree and it's even worse! It also depends on the state of the receptor system you are testing, whether it is active or not.

Hardie: Apropos this point and your question about what is the best drug. I'd just like to remind people of some observations that we made with using acute challenges of either lisuride or apomorphine in patients who had just switched off within 10 minutes or so. We found that lisuride was only effective somewhere on between a half and two thirds of the occasions, whereas apomorphine was universally effective. And I think this may be relevant to the discussion. I just wonder whether Dr. Horowski is correct do dismiss apomorphine as being unlikely to be practicable in the clinical situation and wonder whether anybody else has any comment.

Mereu: From data of our laboratory and other laboratories we observed that the inhibitory effect of a dopamine agonist such as apomorphine or dopamine iontophoretically applied on dopaminergic neurons depends on the basal firing rate of the dopaminergic cell. If the cell is spontaneously active at a very low rate, dopaminergic agonists are more effective in inhibiting activity, whereas if the cell is firing at a high rate dopaminergic agonists are less active. The question is what is the basal activity of the remaining spontaneously active neurons in Parkinson's disease that maybe will reflect the therapeutic efficacy of a dopaminergic agonist?

Horowski: The problem we had with apomorphine constant rate infusions was that the renal threshold was extremely low. Whatever we gave went away through the kidney, and that is why the oral application was not effective. It was not bad absorption and poor bioavailability but it was low renal threshold. Whatever amount of apomorphine you infused went away and little reached the brain. So, you should have given pulses in order to achieve something on the brain receptors. That is the reason why I feel that apomorphine itself is a very bad candidate for constant release forms.

Lees: Following on from the work Richard Hardie described and other observations that also showed that apomorphine was very effective in switching people who are off, on, we started some apomorphine infusions. We have only done one patient, but what I can say is that the patient tolerated the drug very well. We have been able to keep the patient on for long periods, several hours, and had no side effects, provided we blocked with domperidone.

Marsden: What dose was that?

Lees: Up to a maximum of 16 mg in a 12-hour infusion period and up to 3 mg/hour.

Stahl: There are at least two important variables we are talking about here. One has to do with the mode of input, whether it is pulsatile or continuous, and the other has to do with whether it is necessary to stimulate D 1 or D 2 receptors, or both. How about this for a hypothesis? Certainly we know that pulsatile delivery of D 2 agonists sensitizes, whereas continuous delivery tolerates. Is it possible that in early stages of the disease dopamine is stored when taking your oral therapy and is continuously released because of the buffer. Whereas late in the disease you are converting to a pulsatile delivery due to the lack of the buffer, and therefore causing a sensitization? We have emulated some of these things in animal models and have found that D 1 agonists can block the pulsatile-induced sensitization and they can also block the continuous-induced tolerance. Maybe what we need is a D 2 agonist as a background, and the signal when you are turning off is you have lost a little bit of D 1. The experiment might be to take a D 2 agonist and to see whether SK&F can turn you on. And therefore the hypothesis means you need continuous D 2 stimulation and that D 1 is a necessary but not sufficient property of turning back on. Does this make any sense?

Gessa: I think we don't have any evidence that D 1 stimulation is of any positive efficacy in parkinsonian conditions.

Horowski: Answering to Dr. Stahl's comment, I am wondering what system he was testing for the sensitization and tolerance development. To confuse the situation even more, in the same animals we got tolerance to some effects, unchanged response to others, and supersensitivity to other effects. The same holds true for humans. You get tolerance, and that is the beauty of Obeso's and others data with the constant infusion system, that for some responses you get very nice tolerance, so you can achieve much higher plasma levels whilst for the motor response obviously the system remains active or even supersensitive.

Marsden: The question that has been raised by some of those listening here is the significance of these nodules on subcutaneous administration which, it turns out, sound not to be specific for lisuride. The use of the methylester subcutaneously is producing similar nodules. What significance should be attached to these nodules? Are they a practical disadvantage to subcutaneous administration?

Obeso: Getting rid of the nodules will be a major advantage. It will minimize further the problems in the availability of lisuride or other compounds.

Marsden: They clearly are interfering on some occasions with the absorption of the subcutaneously administered agonist, introducing a degree of variability in response. You didn't actually tell us in your talk how often you change the needle.

Obeso: We started by recommending 4–5 days. When these nodules were developing we reduced the interval. It varies now between 2 and 3 days on average. There are one or two patients where we change it every day, but that is an exception.

Chase: It is not clear to me, particularly from the report from the Würzburg group, whether using a plastic catheter actually eliminates the nodules or just makes them better.

Przuntek: That is an experience we made with insulin and not with lisuride. With lisuride we have only tested metal needles.

Marsden: Following this conversation, I wonder whether anybody here has experience of listening to the diabetes doctors and whether the development of nodules is a problem with a constant subcutaneous infusion of insulin for the control of brittle diabetes. Does anybody know the answer to that question?

Parkes: It is not a problem.

Stocchi: Our patients developed red nodules, but we could cope better with them because of stopping the infusion during the night. This relieves the patients a little bit of developing red nodules. Another strategy that seems so far to be effective is to change the site of the needle every morning, and a large percentage of the patients get much better.

Obeso: The needle we have been using until now is not perpendicular to the fatty tissue; going in at a 45° angle produces more mechanical damage.

Chase: I think somebody talked about using implantable pumps. And if you have an implantable pump, then you obviously can't change the locations or lumen or the exit of the catheter. Can anybody comment on whether nodules form when you administer lisuride with an implantable pump?

Yébenes: I use lisuride through an implantable pump in animals, but this is into the central nervous system there are no nodules there!

Marsden: What about alternative delivery sites? We may come back to intracerebro-ventricular in a moment, which is interesting. But I'd very much like to push Roger Duvoisin to tell us about the technology of people having constant feeding by a needle into the stomach which he was mentioning.

Duvoisin: I really don't like to say too much in advance. My colleague Jacob Sage has been largely doing this work and I have been encouraging him to continue. In 6 patients it is rather clear. By the way, the results of the first several patients by Kurlan and Shoulson have just been published in the "Annals of Neurology", the last issue, so I refer you to that. It is the same technology basically. We set out to show that in patients with severe fluctuations who nevertheless have periods where they are near normal, that these would be the ideal candidates. Using a nasoduodenal tube in place and running a pump similar to that Tom Chase showed you, wearing a dopa solution in a jacket, it is possible to smooth out the fluctuations quite dramatically. I would say from what I have seen of Tom's patients about the same magnitude of response. The practicalities of the nasoduodenal tube: One of our patients had a feeding gastrostomy put in several years before. He was not properly treated for his parkinsonism, partly because of severe nausea, and so had no treatment and became so disabled that he couldn't swallow, ending up with a gastrostomy. When we got to him, we were able to treat him with Sinemet and appropriate antiemetics and make him ambulatory and functional. But then he had severe on-off cycling, so we then put a short tube directly into the stomach, allowing the tip of the tube to migrate into the duodenum and then attached the pump.

There is no nasogastric tube sticking out of his nose; this is concealed cosmetically. His wife was able to mix up the dopa solutions every evening; no sterility precautions are necessary. So that eliminates a great difficulty with intravenous infusions. This gentleman has been converted from a disabled individual to a normally functioning person who has virtually no off-periods at this point. Our next step is to produce a direct percutaneous implantation of a nasogastric tube, as is now done for feeding purposes in terminal cancer patients. The gastric radiologists have a technique of directly through a percutaneous trocar implanting this tube into the stomach. We have thought that the next step would be to prepare a loop of small intestine that would be dedicated to L-dopa absorption. For the gastroenterologic surgeons this is a very simple matter. We have a number of dogs on which we have now done this. We take a segment of about 12–15 cm lengths of ileum, rotate it, produce an ostomy, bringing it out through the rectus abdominis muscle which gives it a sphincter effect. It is so placed that the peristaltic activity goes in, not out, so it is the reverse of an ostomy. And you might say we are doing a reverse ileostomy with a blind pouch. If you then introduce tablets, you don't need a pump anymore, you can just put a Sinemet tablet, or better yet, we are now working with the CR III sustained-release tablet, in that pouch, you get sustained blood levels at last in the dog over 24 hours. Severe on-off patients will then be offered this operative procedure and could be given a sustained-release tablet. It may seem dramatic introducing the risks of this surgery, but for these patients who are severely disabled, this may be far preferable and I would think this approach may be life-saving for many of these patients.

Marsden: That is another possible method of introducing continuous stimulation. I know that Dr. Yébenes and Stan Fahn are working around direct intracerebro-ventricular administration, at least in animal experiments. Would either of you like to comment for saying this is a practical proposition in patients?

Yébenes: I think that there may be advantages of direct intracerebral infusion. The main reason may be that you could direct your dopamine agonist to the brain regions where you need the effect, mainly to the basal ganglia. The data that we have so far are very preliminary, but we could increase dopamine concentrations in selective places and we may be able to avoid psychiatric disturbances and some other disturbances related to other dopaminergic systems.

Marsden: That speaks to the science of the technique. I am thinking of the practicality of the technique. What is the longest time brain cancer surgeons have had implantable preparations in the human head without problems?

Fahn: Is there anybody who can answer that specific question about how long they keep these pumps in? My view is that they probably have been able to keep the pumps in for some time as long as the patients are still alive, but I don't have hard data on that. It is clear that brain surgery for treating severe parkinsonism perhaps may be the far extreme. There may be certain advantages of some of the surgical techniques over medical techniques. We have already heard about one from Tokyo, where Narabayashi has treated the dyskinesias from levodopa by doing thalamotomies. Then the patient can take a Sinemet constantly by mouth without dyskinesias, and in big enough dosis so they don't get off-phenomena. One of our limitations in managing the off-phenomena is that we try to prevent dyskinesias, so we try to keep the dose somewhat on a lower side.

Marsden: Perhaps I could bring us to the next topic, that is the complications of this form of therapy, whatever the route and whatever the drug. Again quickly totting up numbers, I think about 60 patients have been presented with subcutaneous lisuride infusions and roughly 30 of them had experienced some form of psychiatric side effect. That overwhelmingly seems to be the major problem concerned. How are we going to stop that?

Quinn: My answer is, I don't know. But perhaps I could make just a couple of comments, because we have had to exercise our minds considerably more on this problem than some other groups. First of all, has it something to do with the disease that we are treating? Not specifically, because we have treated patients with dystonia with oral lisuride, and about 7 out of 42 or so had psychiatric complications. There are also case reports of people developing hallucinations on lisuride given for prolactinoma. So it isn't necessarily just the disease. Is it the route? Some of these other diseases are only treated orally and they can develop problems, too. Has it something to do with previous treatment? That may be a significant factor between our different groups. Our patients tend to have been previously exposed to higher doses of levodopa. I would like to pose another question to people who have been able to use the drug clozapine, which we have not had available in the U.K. Is it really true that you can abolish the psychiatric side effects without making the parkinsonism worse, because that will be a very valuable drug, if it is true.

Marsden: A question has been asked as to whether clozapine really can abolish psychiatric complications of dopamine stimulation in patients with Parkinson's disease without making Parkinson's disease worse. And I think we turn to our German colleagues who have a lot of experience with the use of clozapine.

Przuntek: Our opinion is that we can't improve the patients with clozapine, and that the difference between clozapine and other neuroleptics is not so great as generally has been said. We can also observe, if you have a severe parkinsonian syndrome, that the parkinsonian syndrome deteriorates if the patients take clozapine.

Klawans: I'll talk about psychosis this afternoon. I would like to remind everyone that the normal brain, as far as we know, frequently develops a paranoic-psychotic reaction, not terribly different to this often associated with the abnormal involuntary movements from chronic oral or intravenous use or abuse of amphetamine, methylphenidate, and a whole host of other drugs which have no structural relationship with the ergot derivatives under discussion, but merely are dopamine agonists.

Tolosa: I was surprised to hear from you, Niall Quinn, that you don't seem to think that the disease is contributing to these psychiatric side effects. I think it is probably important to keep in mind that some form of mental deterioration or possible minor dementia, which is not uncommon in Parkinson's disease, has something to do with the psychiatric reactions to several drugs.

Marsden: The question is, of those 30 odd who developed psychiatric complications of subcutaneous lisuride infusions, what proportion had shown psychiatric illness or psychiatric problems before they started the lisuride infusions? In the London experience virtually all our patients had had some degree of psychiatric complications from our previous overtreatment to try and control their motor fluctuations before. José Obeso, I think 10 of your patients had some former psychiatric complications, 10 of 29. How many of those 10 of 29 had had prior psychiatric illness?

Obeso: Two.

Marsden: In Italy, if we accept that hypersexuality may be looked upon by the Anglo-Saxons as a deviation from normal, how many of your 6 patients with psychiatric problems on lisuride had prior difficulties?

Stocchi: Four.

Marsden: Four of 6 in Rome, 2 of 10 in Pamplona, 12 of 12 in London. In Germany?

Przuntek: In Germany also all patients had psychiatric problems before treatment with lisuride. But I believe, it is very essential to adapt the dosage to the need of the patient. And I think it is very important to have a pump with which we can regulate the dosage we need for the day. We see that the therapeutic window is very narrow. If we elevate the dosage only by 0.1 or 0.2 mg/day, then we can see psychiatric side effects. And if we go down a little bit, then we can forget these psychiatric side effects.

Marsden: So that the rough straw poll about that data is that probably two thirds of those who developed psychiatric complications on lisuride had exhibited psychiatric problems before they took lisuride.

Klawans: I think that is an incredibly important part of the parcel because it may be that we are testing lisuride on the world's worst candidates. Even if we assume that lisuride's therapeutic index in relation to psychosis versus overall control of motor fluctuations may not be as good as levodopa, this becomes a very exquisite dose-limiting factor. And if half or two thirds of the patients we are studying with lisuride have this exquisite dose-limiting factor, it is clear what you are going to do with it. Then it may really be a problem of patient selection as far as giving lisuride a fair trial.

Toone: I can refer to a study by Turner et al. from Bartholomew's Hospital in London. They were treating prolactinoma patients with bromocriptine and they reported 10 cases on very low dosages, none of whom had a previous psychiatric illness, who developed psychoses, and the psychoses resolved after the drug was withdrawn.

Spouliady: The question is whether there was a qualitative similarity between the pre-lisuride psychiatric illness and the psychiatric syndromes that developed when the infusion was given. We were just told that there were previous psychiatric illness. I think that is important because we know that Parkinson's disease, even untreated cases, have a number of psychiatric problems which are not necessarily psychotic. The other thing about the psychiatric complications before the lisuride treatment is that I suppose from what I heard this morning, patients who have lisuride are in very advanced stages of the disease and they may have been at completely different stages when they were treated with other drugs earlier on. Perhaps it is not surprising that they may develop a psychiatric complication at the very late stage of the disease.

Marsden: Can I ask Dr. Critchley to respond with his great experience of psychiatric complications of lisuride. Were the nature of the psychiatric complications during lisuride infusions similar to the psychiatric problems that they had prior to these infusions?

Critchley: The answer is yes.

Horowski: I just want to comment on the prolactinoma patients. There are those reports and we have looked through all our files to take them together. They are extremely rare and there are even reports that hallucinations did occur in prolactinoma patients without any drug treatment.

Tolosa: Was there any relation in your experience with the severity of the stage of the disease and the appearance of psychiatric side effects by the Hoehn and Yahr scale? You said, two thirds of the patients with psychiatric side effects had previous history of psychiatric problems. In the other one third or perhaps also for the whole group, was there any relationship with the severity of the illness?

Obeso: Going along with Harold Klawans' comment, there seems to be a certain correlation. You take together Rome's results with, I think, slightly less severe patients, our results with a combination of medium and very severe patients and Würzburg and London with apparently very severe patients. There seems to be a certain correlation with the incidence of psychiatric side effects. The London group being treated with rather high doses of levodopa, probably because they needed it, and with a number of dopamine agonists, our patients being treated with a mixture in between and the Rome group being able to get away without levodopa, there seems to be a correlation between these three stages and the incidence of psychiatric side effects. Probably indeed from the point view of psychiatric side effects we got the wrong group of patients. Why we started with them is because they were the ones who needed this treatment most from the point of view of motor capacity.

Tolosa: It seems to me that this is an important issue. You reported about 6 patients who were treated with lisuride alone. Did any of them have psychiatric side effects?

Stocchi: No, none of the six.

Toone: I think it is important to tease out the psychiatric phenomenology in greater detail, because up to now most of the syndromes have been described as either organic confusional states or psychoses. I suspect that if one looks at this very carefully one will find different sub-groups. I am particularly impressed by the occurrence of sensory perceptual abnormalities in clear consciousness, particularly visual hallucinations, and also changes in other sensory modalities like taste and smell. I wonder whether these effects do have a dopaminergic basis. They are not very common in schizophrenia. They may have more resemblance to the use of hallucinogens like LSD and mescaline. I think it is worthwhile considering whether these have a serotonergic basis.

Yébenes: There is another reason for the simultaneous use of atypical neuroleptic drugs together with dopamine stimulation. There are pharmacokinetic reasons for it. A few years ago we were trying to study the effect of different benzamide derivatives including oxiperomide, tiapride and sulpiride on the complications of L-dopa treatment. We did what we used to do with all patients, that was measuring L-dopa curves after medication, before and after any change in medication. What we found was what we expected, that all of these benzamide derivatives increased the area under the curve of L-dopa in plasma they increased the peak levels of dopa and they decreased the time delay from the appearance of the peak. That may mean that you may not give to the patients such a big amount of dopa. You may obtain better therapeutical results with lower doses of dopa and that may mean that psychiatric complications may be lower.

Bittkau: To answer the question whether on lisuride alone there are some patients who can develop psychoses: We had treated 3 patients with hyperkinesias, and there was one patient who never before had psychotic side effects or psychotic events; she developed slight hallucinations. And we had another young patient with Parkinson syndrome who was not pre-treated with L-dopa and who within one week on lisuride alone developed psychoses.

Rinne: We have to remember that when using dopamine agonists orally, most advanced cases can easily develop psychiatric side effects. I have the same experience with infusions. We have treated two patients, very advanced cases, and both had psychiatric complications. But fortunately they were partially reversible by reducing the dose of lisuride. We have to remember that also very mild cases can have psychiatric complications with oral doses of dopamine agonists. I suggest that also the sensitivity of limbic dopamine receptors might be responsible, at least in some cases.

Tolosa: I would like to make a comment about a personal experience with N-propylnorapomorphine. I think from what I have heard today, lisuride produces the same type of mental alterations that oral apomorphine and N-propylnorapomorphinproduce. I think this argues in favour that those are dopamine-mediated.

Complications in treatment

J Neural Transm (1988) [Suppl] 27: 117–122

Psychiatric side effects during the treatment of Parkinson's disease*

H. L. Klawans

Department of Neurological Sciences Rush-Presbyterian St. Luke's Medical Center,
Chicago, Illinois, U.S.A.

Summary. Dopaminergic agents including both levodopa and direct-acting agonists induce a variety psychiatric side-effects of which psychosis is the most significant. When this occurs early in the course of treatment, there is usually a history of prior psychotic illness. Chronic treatment can, however, elicit psychosis in individuals without such a history. The possible pathogenesis of this is reviewed.

*

The mental status of patients with Parkinson's disease receiving chronic treatment with levodopa and the various dopamine agonists has become increasingly important. In both instances a distinction must be drawn between mental changes related to progression of disease and/or aging and those mental changes that are clearly pharmacologically induced. Although dementia was not originally described as part of the clinical spectrum of Parkinson's disease, the association of dementia with this disorder was well documented before the advent of the levodopa era. The well defined incidence of dementia in Parkinson's disease prior to levodopa therapy, the observation that most patients who became severely demented on chronic levodopa were already experiencing mild dementia prior to treatment, the failure of levodopa or anticholinergic withdrawal to reverse dementia and the pathologic evidence of a more widespread degenerative process involving far more CNS areas than the substantia nigra alone all suggest that the mental changes characterized as dementia are not related to the pharmacologic treatment of the disorder, but are probably associated with advancement and progression of the basic disorder. This is not true of the psychotic behavior seen during active dopaminergic therapy. Since the psychiatric side effects have been best studied in relationship to levodopa, levodopa induced psychosis will be discussed as a model of psychosis in Parkinson's patients receiving dopaminergic therapy.

---

* This work is supported in part by grants from the United Parkinson Foundation, Chicago, Illinois, and the Boothroyd Foundation, Chicago, Illinois.

Levodopa-induced psychosis occurs in two separate clinical situations: (1) early in the course of treatment, when it is seen in patients with a previous history of mental disease, and (2) after several years of treatment, when it occurs in patients with no previous history of psychiatric disorders.

When levodopa-induced psychosis occurs early in the course of levodopa therapy it usually happens within a few weeks after the levodopa regimen has started. Almost invariably the patient has had a past history of severe psychiatric disorders, and most often the diagnosis has been schizophrenia. The manifestations of levodopa-induced psychoses in such patients can be quite variable. They often resemble the previous psychotic manifestations the patient had, although most of these patients exhibit some element of paranoid thinking after taking levodopa even if such thinking has not been present previously. These patients frequently have a formal thought disorder as well as all other manifestations of schizophrenia.

This levodopa-induced exacerbation of pre-existing schizophrenia has been observed in two patient populations. The first group includes those patients who were receiving neuroleptic treatment for their psychoses and, upon developing drug-induced parkinsonism, were given levodopa. This phenomenon is probably analogous to the exacerbation of schizophrenic symptoms by amphetamine and methylphenidate, both of which act by increasing the activity of dopamine within the striatum. These psychiatric symptoms exacerbated by levodopa are invariably reversible once the levodopa is discontinued. The high incidence of levodopa-induced exacerbation of pre-existing schizophrenia in this group of patients, however, serves as an absolute contraindication to prescribing levodopa for drug-induced parkinsonism.

The second group of those patients with a past history of psychosis who were psychiatrically asymptomatic when levodopa therapy was initiated. After being treated with levodopa for idiopathic parkinsonism, the psychosis reappears. These psychotic symptoms, again, are almost always self-limited and usually in days to weeks reverse once levodopa is withdrawn. A past history of schizophrenia should be considered a contraindication to the use of levodopa in a parkinsonian patient, since occasionally patients in this group will develop a psychosis which will last for several months after the levodopa is discontinued.

Levodopa is also able to induce a psychosis in parkinsonian patients without any past history of psychiatric disorder. While cerebral cortical atrophy accompanied by dementia is common in parkinsonian patients and may predispose to this complication, levodopa-induced psychosis also occurs in parkinsonian patients without any evidence of dementia. This psychosis usually occurs after several years of levodopa therapy and is definitely related to two other levodopa-induced mental alterations: dreams and hallucinations (Moskovitz et al., 1978).

Three types of dream alterations occur in chronic parkinsonian patients receiving levodopa over a long period of time. They occur in at least onethird of all patients taking levodopa for at least two years. They may be classified as vivid dreams, night terrors and nightmares.

Vivid dreams

These levodopa-related dreams are qualitatively quite vivid, seemingly real, temporally condensed, internally organized and coherent, often affectively neutral, with a frequent theme of persons and events from the dreamer's remote past. These dreams are qualitatively different from the patients previous dream experiences and are by far the most common type of levodopa-induced dreams.

Night terrors

Since patients are always amnesic during these experiences, they are reported by other members of the family. Typically the patient screams, calls out and thrashes around during sleep. He may awaken screaming but forget why he woke up. About 5 percent of chronic levodopa patients have such experiences. In yet another subgroup of patients, the experiences, although accompanied by thrashing and crying out, are characterized by unexplained laughter. Perhaps unfortunately, patients are also amnesic for these events.

Nightmares

These are classic, frightening, often paranoid nightmares, and − like the vivid dreams described above − are considered by the patients to be distinctly different from other dreams. About 5 percent of chronic levodopa patients report nightmares.

About one half of the patients on levodopa for two or more years develop hallucinations. The hallucinatory phenomena frequently are stereotyped in each patient; they are often nocturnal, nonthreatening and recurrent. The hallucinations are predominantly visual (at times with a secondary auditory component) and are superimposed on a clear sensorium. They usually conform to boundaries imposed by actual concurrent sensory input and often concern individuals and experiences that were significant in the patient's past. At times, such hallucinations blend indistinguishably with dream phenomena possessing similar themes. At least two thirds of these hallucinations are associated with levodopa-induced dreams. As hallucinations persist, it may become more difficult for the patient to distinguish the phenomenon as unreal. In some cases, the apparent presence of unwelcome of unfamiliar persons in the home is frightening, and patients may bother family members with complaints about these intruders.

Psychosis is the third and most important psychiatric side effect of chronic levodopa therapy. The majority of these psychoses are paranoid in nature. Most commonly they manifest as a pure paranoid delusional system, superimposed on a clear sensorium, with no other qualities of thought disorder present. About 3 percent of the patients who have been treated with levodopa for two or more years experience such psychoses have associated with confusional elements. All levodopa-induced psychotic states are associated with vivid dreams and/or hallucinations. Levodopa-induced confusional psychoses have been found to be usually associated with pre-existing paranoid nonconfusional psychoses.

Overall, then, this suggests a progression of medication-induced psychiatric symptomatology from striking dreams, to dreams plus hallucinatory experiences, to a pure paranoid delusional system in addition to the former manifestations, and, finally, to a confusional state superimposed on the whole. The onset of psychiatric symptoms varies in these patients from insidious to acute. Although the onset of any of these phenomena may be occasionally triggered by an increase in antiparkinsonian therapy, usually there has been no associated alteration in maintenance therapy.

In a more recent study, Nausieda et al. studied sleep alterations occurring during the course of chronic levodopa therapy (1982). They found that sleep disruption was a common complaint in levodopa-treated parkinsonian patients. A survey of 100 parkinsonian patients revealed prominent sleep complaints in 74%. Sleep complaints were unrelated to patient age and the duration of disease but increased in prevalence with long periods of levodopa therapy. Sleep abnormalities tended to increase in severity with continued treatment and insomnia tended to be followed by daytime somnolence, altered dream events, and episodic nocturnal vocalization and myoclonus. While dyskinetic side effects and on-off syndrome were encountered in patients with and without sleep complaints, 98% of patients experiencing psychiatric side effects also reported sleep disruption. The authors suggested that sleep-related symptoms constitute an early stage of levodopa-induced dopaminergic psychiatric toxicity in the parkinsonian population.

The exact pathophysiologic basis of all of these abnormalities remains unclear. Many of the patients with levodopa-induced psychosis and sleep disruptions also manifest levodopa-induced myoclonus (Nausieda et al., 1982).

The myoclonic jerks were not restricted to any specific stage of sleep. Patients with myoclonic movements during the waking state [levodopa-induced myoclonus (Klawans et al., 1975)] universally had similar events during sleep, suggesting they were identical events. Subjectively, the episodes are similar to those previously described under the heading "nocturnal myoclonus" (Coleman et al., 1980). Previous studies had not associated "nocturnal myoclonus" with sleep architectural abnormalities of the type described in our patients, nor has an association been made with altered dream events. Unlike the movements previously reported, myoclonic movements in the parkinsonian patient population are not associated with spontaneous awakening and are usually reported by the bed partner (who is frequently awakened by the movements). These features suggest that myoclonic movements during sleep in parkinsonian patients are probably distinct from those previously reported in other clinical settings.

While the sleep alterations that occur in patients receiving chronic levodopa are of interest as previously unrecognized side effects of therapy, of greater interest is the association of the symptom complex with levodopa-induced hallucinosis. While earlier studies have identified altered dreaming as an early feature of psychiatric toxicity (Moskovitz et al., 1978; Sharf et al., 1978) other components of sleep disruption were not appreciated. The study of Nausieda et al. showed that psychiatric reaction usually occurred in the presence of sleep disruption (43 of 44 patients), and conversely, prominent sleep disruption was found to be a characteristic of patients with hallucinatory episodes. Using cumulative sleep disruption score as an index of severity, patients with hallucinosis had an average score of 8 (range, 0–16) while patients without hallucinations demonstrated an average score of 2 (range, 0–6). These authors interpreted these data as indicating that levodopa-induced hallucinosis was not a monosymptomatic disorder but occurs in the context of a more widespread disruption of sleep-wake behavior. Myoclonic activity during sleep appeared to be the best predictor of waking hallucinosis. Since myoclonic movements appear to be late complaints and occur only in the context of multiple sleep alterations, these data supported the view that hallucinosis is a late event in patients with progressive sleep derangement.

It is most tempting to suggest that all of these problems relate to dopaminergic agonism or changes in the dopamine receptors. The answer might not be that simple.

A serotoninergic basis for levodopa-induced hallucinosis has been raised on the basis of postmortem neurochemical studies of patients with this side effect, which demonstrated significantly lowered levels of serotonin in the brain stem (Birkmayer et al., 1974; Birkmayer and Riederer, 1975). It has been previously suggested that L-tryptophan migth improve psychiatric symptoms in patients receiving levodopa, though the results of such treatment have not been conclusive (Miller and Nieburg, 1974; Beasley et al., 1980). Hallucinosis appears to correlate well with the appearance of myoclonic activity. The myoclonic disorder in levodopa-treated patients has been found to respond to the administration of methysergide (Sansert) (Klawans et al., 1975). Myoclonus has also been reported in patients with Down's syndrome receiving L-5-hydroxytryptophan (l-5-HTP) (Coleman, 1971). These observations suggest that overactivity in a central serotonergic system mediates myoclonus in some settings, and may underly myoclonus in parkinsonian patients receiving levodopa chronically. Since there is a correlation between myoclonus and hallucinosis/psychoses it is possible that serotonergic mechanisms may also play a role in the latter.

The pathophysiology of agonist induced psychiatric changes is also unclear but the possibility that they are not purely dopaminergic must be kept in mind.

References

Beasley B, Nutt J, Chase T (1980) Treatment with tryptophan of levodopa associated psychiatric disturbances. Arch Neurol 37: 155–156

Birkmayer W, Danielczyk W, Neumayer E, Riederer P (1974) Nucleus ruber and L-dopa psychosis: biochemical and post-mortem findings. J Neural Transm 35: 93–116

Birkmayer W, Riederer P (1975) Responsibility of extrastriatal areas for the appearance of psychotic symptoms. J Neural Transm 35: 175–182

Coleman M (1971) Infantile spasms associated with 5-hydroxytryptophan administration in patients with Down's syndrome. Neurology 21: 911–919

Coleman RM, Pollak C, Weitzman E (1980) Periodic movements in sleep (nocturnal myoclonus): relation to sleep disorders. Ann Neurol 8: 416–421

Klawans HL, Goetz C, Bergen D (1975) Levodopa-induced myoclonus. Arch Neurol 32: 331–334

Miller E, Nieburg H (1974) L-tryptophan in the treatment of levodopa-induced psychiatric disorders. Dis Nerv Syst 35: 20–23

Moskovitz C, Moses H, Klawans HL (1978) Levodopa-induced psychosis: a kindling phenomenon. Am J Psych 135: 669–675

Nausieda PA, Weiner WJ, Kaplan LR, Weber S, Klawans HL (1982) Sleep disruption in the course of chronic levodopa therapy: an early feature of levodopa psychosis. Clin Neuropharmacol 5: 183–194

Sharf B, Moskovitz C, Lupton M, Klawans HL (1978) Dream phenomena induced by chronic levodopa therapy. J Neural Transm 43: 143–151

Author's address: H. L. Klawans, M.D., Department of Neurological Sciences, Rush-Presbyterian St. Luke's Medical Center, 1725 W. Harrison Street, Chicago, IL 60612, U.S.A.

J Neural Transm (1988) [Suppl] 27: 123–132

Applications of new drug delivery technologies to Parkinson's disease and dopaminergic agents

S. M. Stahl

Neuroscience Research Centre, Merck Sharp & Dohme Research Laboratories, Harlow, Essex, and Institute of Psychiatry, University of London, London, U.K.

Summary. Recent advances in drug delivery technology are creating novel therapeutic approaches to the treatment of Parkinson's disease with levodopa and dopamine agonists. This article reviews those technologies which can be applied to Parkinson's disease, both for targetting the central nervous system with drugs, as well as for matching the appropriate rate controlled delivery with therapeutic needs. In particular, the possibility exists for eliminating erratic highs and lows of drug delivery to the brain, and to substitute rate controlled, constant drug delivery. Clinical investigations now in progress suggest that new technologies which deliver constant dopaminergic stimulation to patients with Parkinson's disease may not only eliminate the unpredictable swings in therapeutic efficacy in Parkinson patients with the "on/off" effect, but may even have a role in the future in preventing such fluctuations from developing in patients chronically treated with dopaminergic therapies.

Introduction

Recent clinical investigations suggest that providing constant dopaminergic stimulation may be a useful therapeutic approach to the management of Parkinson's disease (Nutt et al., 1984; Obeso et al., 1986 a). This approach appears to have the most potential for treating the "on/off" effect and the "wearing off" effect, as well as for treating more moderately ill patients prophylactically so that such long-term complications will not develop (Stahl, 1985). Although early conceptual work with intravenous levodopa infusions (Nutt et al., 1984; Quinn et al., 1982), as well as with intravenous (Obeso et al., 1986 b) and subcutaneous (Obeso et al., 1986 a) infusion of dopaminergic agonists such as lisuride, have attested to the theoretical value of this approach, many practical problems have limited the widespread applications of it to the treatment of Parkinson's disease. Here we review some of the advances in modern drug delivery technologies which have recently become available for clinical applications, and which might be usefully applied to the therapeutic goal of providing constant dopaminergic stimulation in Parkinson's disease.

Targetted delivery to the central nervous system

It has long been recognized that getting a drug to its intended therapeutic site of action within the central nervous system (CNS) can be a formidable task. *Chemical* barriers such as enzymes and membranes not allowing charged hydrophilic molecules to penetrate the brain are coupled with *physical* barriers such as blood capillaries, glial cells, astrocytes, and the extracellular space, including cerebrospinal fluid (Bates, 1984). The techniques used to overcome these barriers, to allow molecules to penetrate the CNS, can also be categorized into *chemical* approaches and *physical* approaches (Bates, 1984; Gardner, 1985) (Table 1).

The usual approach to delivering a drug into the CNS is to use a *chemical* approach whereby the octanol: water partition coefficient (Log P), is sufficiently high to allow the molecule to transfer passively into the CNS. Potent drugs will be active at the CNS concentrations they reach by such passive diffusion. Of course, such drugs will also penetrate other tissues, and no selectivity for the brain is conferred by this approach.

Another chemical approach which attempts to surmount the blood brain barrier utilizes liposomes as potential delivery systems to the brain (Henderson, 1983). Liposomes are tiny spheres, much smaller than body cells, that are artificially manufactured from fatty molecules. They are easy to prepare, are biodegradable, and can be made to entrap a variety of drugs either in their hollow interiors or within their lipid structure. Unfortunately they tend to be unstable once administered to the body, and are rapidly degraded by the reticuloendothelial system. Nevertheless, the possibility of creating such a "Trojan Horse" to sneak peptides or other difficult-to-deliver molecules into the brain remains an exciting possibility.

Chemists can also exploit transport pumps in order to deliver drugs to the brain. Drug delivery efficiency is increased by chemically modifying the active drug to a form, called a prodrug, that the body absorbs and converts into the active compound (Gardner, 1985; Henderson, 1983). The chemical modification

Table 1. Drug targetting of the CNS

Chemical approaches
    passive diffusion (partition coefficient)
    liposomes
    prodrugs
    transport pumps
Physical approaches
    Ommaya reservoir
    intraventricular catheters plus implantable pumps
    nasal aerosols
    buccal (chewing)
    pulmonary (smoking/inhaling)

---

provides the active molecule with a chemical "mask" that the body removes by way of an enzyme or some other means once the agent has penetrated the brain. Since most of the prodrug reaches the target, less of it is available to produce undesirable side effects in peripheral tissues. This approach has sometimes been called "site selective prodrug reconversion" (Gardner, 1985; Henderson, 1983). One example of this prodrug approach is, of course, the development of levodopa, which after all is merely a prodrug of dopamine. Levodopa exploits the active transport mechanism for amino acid uptake into the brain, and once transported, is enzymatically converted to dopamine. Another approach along this same line is the synthesis of a prodrug which combines dopamine with a molecular mask derived from the fat soluble vitamin niacin. This "modified dopamine" is taken up into the brain where it is converted to dopamine itself at a rate slow enough to meet the brain's needs (Bodor and Simpkins, 1983).

Physical approaches (Table 1) can also be utilized in order to deliver drugs into the CNS, but at the present time they are rather invasive. One such method is the implantation of a silastic catheter into the lateral ventricle of a patient, and attaching this to a subcutaneous reservoir called an Ommaya reservoir (Ommaya, 1984). Drugs are injected into the reservoir and squirted into the ventricle when the reservoir is compressed. More recently, these silastic catheters have been connected to more sophisticated subcutaneously implanted pumps such as the Infusaid or Metal Bellows pump (Ensminger et al., 1981; Harbaugh et al., 1984.

There are several examples of these types of implantable pumps, more commonly used in the delivery of chemotherapy (Ensminger et al., 1981) and insulin (Mecklenburg et al., 1985). Recently, these pumps have been used to deliver the central cholinergic agonist bethanechol, which cannot penetrate the blood brain barrier, for the treatment of Alzheimer's disease (Harbaugh et al., 1984). The Infusaid infusion pump is a disc-shaped implantable pump usually placed under the skin of the abdomen, containing a compartment filled with a liquid fluorocarbon propellant that presses against a collapsible reservoir containing the desired medication. Body heat causes the propellant to vaporize and expand, compressing the reservoir and forcing the medication out through a catheter. At intervals the drug reservoir is refilled from the outside by a hypodermic injection, which also recompresses the fluorocarbon to a liquid.

Other more sophisticated pumps include Medtronic's battery powered implantable system which also contains a refillable drug reservoir, a motor driven pump, and even a micro electronic control element that can be reprogrammed from outside the body, including the possibility of utilizing a telephone from a physicians office, in order to vary the pump's delivery rate. Many other implantable pumps exist, but are too numerous to review here.

Another approach to the delivery of drugs to the CNS which is earlier in development is the possibility that installation of drugs through a rate controlled system placed on the conjunctiva may deliver drugs through the nasal lacrimal gland, where it would be absorbed by the nasal membranes, and then be delivered directly through the boney cribriform plate, into the brain. Nasal delivery systems, including nasal aerosol sprays (Salzman et al., 1985), are currently under development for the delivery of several CNS active peptides such as vasopressin, somatostatin, and others. The buccal membrane surface can deliver drugs through a resin which is chewed, allowing for rapid uptake into the vascular supply of the face and delivery of the drug into the brain. This has recently been applied to the delivery of nicotine in chewing gum to assist in smoking cessation. Finally, inhalation via the pulmonary arterioles is one of the most effective ways of delivering drugs rapidly to the brain. Inhalation has long been known by the layman to be a most effective drug delivery device for drugs of abuse such as cigarettes, cocaine, heroin, and others.

Rate controlled delivery of drugs to the CNS

It is not only necessary to get drugs into the brain, but to get drugs there when they are needed and at the required rate (Goldman, 1982; Urquhart et al., 1984). Chemists and pharmacologists are generally so concerned with getting drugs into the brain, that they end up delivering too much at times, and sometimes quite literally make drugs "bash the brain". The ideal system, which largely does not exist, should achieve its therapeutic effect immediately upon the start of treatment, remain at a constant non-toxic level during therapy, and end instantaneously and completely. What happens in practice, however, is usually very different and can be quite variable depending on how the drug is formulated. The graphs in Fig. 1 show the concentrations of levodopa in the bloodstream of three theoretical patients as they change with time after taking the same amount of drug in three different formulations. With formulation A, which is absorbed rapidly, drug concentration peaks early but reaches a toxic level, causing undesirable side effects. Formula B is safer but its slow absorption rate both delays the start and shortens the duration of effective therapy. The drug in formulation C enters the blood slowly, but it never reaches an effective

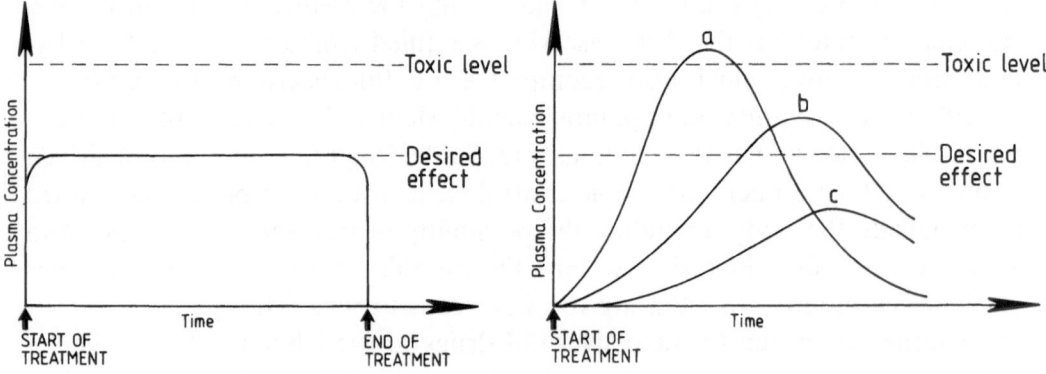

Fig. 1

concentration. Even though formulation B appears to be the best compromise, taking repeated doses produces a rollercoaster pattern of highs and lows in the blood concentration. Whether such patterns contribute to the production or the maintenance of the on/off effect in Parkinson's disease, is of course a matter of intense current investigation.

An example where rate controlled delivery of a CNS active drug has been shown to be important is the recent development of scopolamine in a transdermal system (Shaw and Chandrasekaran, 1978). For this drug, its effects in preventing and treating motion sickness occur at plasma levels below those of its adverse effects such as dry mouth, tachycardia and confusion. A conveniently applied transdermal system allows nearly immediate delivery of scopolamine through a small patch placed behind the ear on the skin and which lasts up to three days.

A second example demonstrating the value of rate controlled systems can be taken from the regulatory control of growth hormone (GH) on synthesis of two liver proteins: hepatic prolactin/GH receptors, and major urinary protein (MUP) (Norstedt and Palmiter, 1984). Experimental data suggest that differences in the secretory rhythm of GH are responsible for the sex-dependent liver function of mice. Prolactin/GH receptors are predominant in females whereas MUP is predominant in males. The secretory pattern of GH is continuous in females, but more pulsatile in males. Mice genetically deficient in GH, synthesize neither liver protein. Continuous delivery of GH by osmotic minipumps in these latter animals reproduces the normal female pattern of GH delivery, and also induces the synthesis of prolactin/GH receptors (normal female response). Pulsatile delivery, mimicked by injections, reproduces the normal male pattern of GH delivery, and also induces the synthesis of MUP (normal male response). These results suggest that induction of MUP messenger RNA requires pulsatile occupancy of GH receptors, which is achieved naturally in males or by pulsatile injections of GH, whereas chronic occupancy of GH receptors is inhibitory. In contrast, induction of prolactin/GH receptors requires chronic stimulation of GH receptors, which is approximated in normal female mice or results from increased GH levels in mice undergoing infusions from minipumps. These provocative observations suggest that *continuous* delivery of drugs might also have dramatically different long-term effects from those arising from *intermittent* peaks-and-valleys delivery of drugs. We have already mentioned this phenomenon may help to explain the production or the maintenance of the on/off effect in Parkinson's disease, and demonstrates the potential value of therapeutic trials of constant dopaminergic stimulation utilizing modern drug delivery technology in Parkinsonian patients.

Costall and co-workers (1983, 1984) have studied intermittent versus continuous dopaminergic stimulation in rats, and have shown that large differences in locomotor behavior can be seen when dopaminergic agonists are given intermittently as compared to continuously. Also, the accoustic startle response of mice, which is potentiated by the dopaminergic drug amphetamine, is *further sensitized* when amphetamine is given by *intermittent* administration, but there is *no* sensitization when amphetamine is given by continuous infusion (Kokkinidis, 1984). Finally, preliminary investigations from our own laboratories suggest that intermittent delivery of a selective dopamine-2 agonist produces tolerance to motor hyperactivity in rats during the day, whereas intermittent delivery causes a sensitization of the same behaviour (Martin-Iverson et al., 1987).

Drug delivery devices

The number of drug delivery devices which are available for practical administration of drugs in man is exploding and cannot be reviewed comprehensively here. However, a few representative examples from the drug delivery armamentarium currently available will be mentioned (Table 2).

One of the most commonly used drug delivery devices has been in therapeutic application for many years. This is a non-implantable intravenous infusion device which stands at the patient's bedside, and is commonly used to control intravenous infusions of medications for use in intensive care units, such as antihypertensive agents, and agents for the treatment of shock. Adaptations of these bedside systems have been made which allow the patient to have some degree of control over the medication administered. For example, one adaptation allows patients recovering from surgery to inject themselves with an analgesic as needed. At the touch of a button, the computerized pump ejects a small amount of morphine or other pain-killer within the limits set by a physician. Patients who receive conventional injections of narcotic analgesics are often over-medicated early on and then find themselves in pain for several hours as their injections wear off.

Miniaturized wearable models, strapped to the body or carried in clothing, are also made as adaptations to the first generation devices used at the bedside only. These wearable models deliver a continuous, often adjustable, supply of medicine by way of a catheter surgically implanted in a blood vessel. These

Table 2. Examples of drug delivery devices

Implantable infusion devices
    osmotic minipumps
    polymers
    Ommaya reservoir
    Infusaid pumps (fluorocarbon propellant)
    Medtronics pump (battery driven and programmable)
Wearable infusion devices
    Travenol infusion (balloon filled reservoir)
    Tecensa, Hoechst pumps (battery driven)
    transdermal delivery systems
Ingestible infusion devices
    osmotic minipumps
    polymers

---

models also have been used for the delivery of insulin and are under investigation as drug delivery devices for subcutaneous administration of dopamine agonists such as lisuride (Obeso et al., 1986 a).

A number of implantable systems are also under investigation, although they are currently used predominantly in experimental animals and not in man. The best known of these is the osmotic minipump, also called the Alzet minipump. This subcutaneously implantable pump, about the size of an antibiotic capsule, can deliver a continuous controlled flow of almost any active agent into the body for as long as one month. Its semi-permeable membrane outer shell encapsulates a layer of osmotic agent in contact with a collapsible reservoir filled with the drug solution. Water passing inward through the membrane presses against the reservoir, slowly compressing it and forcing a constant flow of drug solution from the delivery opening. A catheter may be used to direct a drug from a distant implant site of this pump to a precise target, such as intravenously, or even into select areas of the brain.

More primitive forms of drug delivery device include implantable beads which are made of biodegradable polymers, impregnated with drug, which slowly erode and allow release of the drug from the bound polymer. One adaptation of this methodology, which is under clinical investigation, is the subcutaneous implantation of the narcotic antagonist naltrexone in heroin addicts. Another adaptation is the use of silicone-rubber tubes filled with a contraceptive hormone. Such devices can be surgically implanted under the skin of a woman, and the tubes will release the hormone for up to five years. Clinical trials, particularly in less developed countries, have shown this approach to be highly effective as a contraceptive. In the future, such polymers may be impregnated with drugs combined with small magnetic stainless steel beads. In such a case, when an oscillating magnetic field is applied, the beads move, deforming the polymer matrix and quite literally squeezing out the drug. Such a device would be useful for patients who require an on-demand delivery of an implanted drug, and these patients could use a wristwatch-sized magnetic actuator to boost their drug requirements when needed.

One of the most elegant drug delivery systems is the transdermal drug delivery device (Goldman, 1982; Shaw and Chandrasekaran, 1978). These systems are simply skin patches which use the skin as a controlled route of entry to the systemic circulation. The transdermal route is especially attractive for potent medications which require rate-controlled input, and which have a large first pass effect. Delivery of the drug topically is a form of *parenteral* administration, analogous to an intravenous infusion. Transdermal delivery devices are thin, multilayered discs containing a reservoir of the drug, and an adhesive coated polymer membrane that passes the drug at a controlled rate for a period of time up to one week (Goldman, 1982; Shaw and Chandrasekaran, 1978). These discs are often about the size of a small coin, and are affixed to various areas of the skin. Transdermal devices eliminate peaks of drug delivery (e.g. Fig. 1), and maintain a very low but – for potent compounds – effective blood level. Current applications of transdermal delivery devices include the scopol-amine patch for motion sickness, transdermal nitroglycerine for the treatment of angina pectoris, clonidine for the treatment of hypertension, and estradiol for the treatment of menopausal symptoms (Goldman, 1982; Shaw and Chandrasekaran, 1978).

Various forms of osmotic pills are available for orally administered drugs. For example, osmotic systems analogous to the osmotic and Alzet minipumps, can be manufactured and swallowed rather than implanted subcutaneously. Oral osmotic systems attempt to combat the significant disadvantages associated with conventional tablets, capsules, or liquids, which expose the body to a roller-coaster pattern of drug concentrations in the blood that may produce inadequate therapy or severe side-effects (Fig. 1). The best known of these devices is the Oros tablet, which resembles a conventional tablet in size but uses osmosis to provide the engine for continuous controlled release for as long as 24 hours, about the time it takes to pass through the gastrointestinal tract. A semi-permeable membrane of a synthetic polymer, pierced with a single laser-drilled hole, surrounds a core of solid drug and an osmotic agent. After the tablet is swallowed, water enters through the membrane by osmosis and dissolves the drug. Drug solution is then forced out through a hole at a constant rate that depends on the membrane's ability to control the admission of water. These osmotic devices have important advantages. The drug release rate is unaffected by changes in agitation, acidity or alkalinity within the gastrointestinal tract. The intestinal membranes receive the medication already in solution and ready for absorption, possibly reducing the chance of irritation and injury. Thus, the system retains the precision of dosing, convenience of administration, and the chemical stability of a drug in solid form while offering the advantage of liquid medication.

Other oral sustained release systems include the use of various polymers which slowly decay and release the drug in a programmed and sustained manner. These latter types of devices are currently under development for the rate-controlled administration of levodopa. Madopar HBS is a hydrodynamically balanced system which is retained in the upper gastrointestinal tract, and which is retained in the upper gastrointestinal tract, and which slowly releases the drug from a matrix. Sinemet CR is a controlled-release type of polymer which also allows the drug to be released is a slow continuous manner.

Conclusion

Drug delivery technologies are ushering clinicians into a new era of therapeutics in which medications for CNS disorders can be administered and controlled as never before. Applications of these concepts to the prevention or treatment of the on/off effect in Parkinson's disease in an exciting area of current research effort, and holds considerable promise for improving the therapy of this disorder.

References

Bates IP (1984) The blood brain barrier and central nervous system penetration. Pharm J 30: 265–268

Bodor N, Simpkins JW (1983) Redox delivery system for brain-specific sustained release of dopamine. Science 221: 65–67

Costall B, Domeney AM, Naylor RJ (1983) A comparison of the behavioural consequences of chronic stimulation of dopamine receptors in the nucleus accumbens of rat brain effected by a continuous infusion or by single daily injections. Naunyn Schmiedebergs Arch Pharmacol 324: 27–33

Costall B, Domeney AM, Naylor RJ (1984) Locomotor hyperactivity caused by dopamine infusion into the nucleus accumbens of rat brain: specificity of action. Psychopharmacology 82: 174–180

Ensminger W, Niederhuber J, Dakhil S, Thrall J, Wheeler R (1981) Totally implanted drug delivery system for hepatic arterial chemotherapy. Cancer Treat Rep 65: 393–400

Gardner CR (1985) Chemical approaches to drug delivery to the central nervous system. Psychopharmacol Bull 21: 657–662

Goldman P (1982) Rate-controlled drug delivery. N Eng J Med 307: 286–290

Harbaugh RE, Roberts DW, Coombs DW, Saunders RL, Reeder TM (1984) Preliminary report: intracranial cholinergic drug infusion in patients with Alzheimer's disease. Neurosurgery 15: 514–518

Henderson NL (1983) Recent advances in drug delivery systems technology. In: Allen RC (ed) Annual reports in medicinal chemistry. Academic Press, New York, pp 275–284

Kokkinidis L (1984) Effects of chronic intermittent and continuous amphetamine administration on acoustic startle. Pharmacol Biochem Behav 20: 367–371

Martin-Iverson M, Stahl SM, Iversen SD (1987) Factors determining the behavioural consequences of continuous treatment with 4-propyl-9-hydroxynaphthoxazine, a selective dopamine D2 agonist. In: Rose FC (ed) Parkinson's disease. J Libbey, London pp 169–177

Mecklenburg RS, Benson EA, Benson JW, Blumenstein BA, Fredlund PM, Guinn TS, Metz RJ, Nielsen RL (1985) Longterm metabolic control with insulin pump therapy. N Eng J Med 313: 465–468

Norstedt G, Palmiter R (1984) Secretory rhythm of growth hormone regulate sexual differentiation of mouse liver. Cell 36: 805–812

Nutt JG, Woodward WR, Hammerstad JP, Carter JH, Anderson JL (1984) The "on-off" phenomenon in Parkinson's disease: relation to levodopa absorption and transport. N Engl J Med 310: 483–488

Obeso JA, Luquin MR, Martinez-Lage JM (1986 a) Lisuride infusion pump: a device for the treatment of motor fluctuation in Parkinson's disease. Lancet 1: 467–470

Obeso JA, Luquin MR, Martinez-Lage JM (1986 b) Intravenous lisuride corrects oscillations of motor performance in Parkinson's disease. Ann Neurol 19: 31–35

Ommaya AK (1984) Implantable devices for chronic access and drug delivery to the central nervous system. Cancer Drug Deliv 1: 169–179

Quinn N, Marsden CD, Parkes JD (1982) Complicated response to intravenous infusion of levodopa. Lancet 2: 412–415

Salzman R, Manson JE, Griffing GT, Kimmerle R, Ruderman N, McCall A, Stoltz EI, Mullin C, Small D, Armstrong J, Melby JC (1985) Intranasal aerosolized insulin: mixed-meal studies and long term use in Type I diabetes. N Engl J Med 312: 1078–1084

Shaw JE, Chandrasekaran SK (1978) Controlled topical delivery of drugs for systemic action. Drug Metab Rev 8: 223–233

Stahl SM (1985) New drug delivery systems – a new approach to Parkinson's disease. In: Marsden CD, Trimble M (eds) Future trends in the treatment of Parkinson's disease and epilepsy. Franklin Scientific Projects Publications, London, pp 16–18

Urquhart J, Fara JW, Willis KL (1984) Rate-controlled delivery systems in drug and hormone research. Ann Rev Pharmacol Toxicol 24: 199–236

Author's address: Dr. S. M. Stahl, Merck Sharp & Dohme Research Laboratories, Terlings Park, Eastwick Road, Harlow, Essex CM20 2QR, U.K.

J Neural Transm (1988) [Suppl] 27: 133–138

Discussion

Motor fluctuations and dyskinesias

Poewe: I'd like to make a comment based on observations we made together with Andrew Lees and Gerald Stern in patients with levodopa-induced dystonia. I agree that chorea is the characteristic type of involuntary movements in situations of peak-dose effects, whereas dystonia is the characteristic type of dyskinesia in off-periods. But what we have also seen is that there is a topographical aspect to this. If you challenge patients with off-period dystonic foot posturing with levodopa for example, as they turn on you get not only limb chorea but also frequently orofacial dystonia.

Tolosa: More than 10 years ago I described for the first time dyskinesias starting during the on-period, that they would disappear during most of the on-period, and that they would reappear at the end of the on-period. Later on Manfred Muenter and Yves Agid described it and gave them different names, diphasic dyskinesias and so forth.

Gershanik: We are presently engaged with Dr. Obeso in analysing a large population of levodopa-treated parkinsonian patients trying to see if we can find out a topographic difference concerning dystonic phenomena and dyskinesias. It is our impression that most of the peak-of-dose dyskinesias are axial or facial or truncal, whereas most of the diphasic dystonias are either in the limbs or generalized. We still have not enough numbers analysed but we believe we are going to have the results by the middle of this year.

Quinn: I have got one patient who has got early-morning blepharospasm, and his eyes are completely shut until his first dose of levodopa of the day works. And very interesting, he doesn't get it again when he turns off later during the day; it is exclusively an early-morning problem.

Melamed: There is one more peculiar thing about this early-morning dystonia problem. Something like 50% of the patients get dystonia while they are still in bed, and the other 50% get it only when they attempt the first steps.

Duvoisin: Those rare patients, who have dystonias as a manifestation of Parkinson's disease, sometimes preceding it by years, generally have an exacerbation on levodopa. Is this dystonia the same as the end-of-dose dystonia?

Fahn: We should keep in mind the different kinds of dystonia seen in parkinsonism. What you just described, Roger, is, dystonia as part of the disease. That kind of dystonia may get better with levodopa or it may not. We have already heard about the off-dystonias. I'd like to make a comment about an observation on that dystonia. It is frequently painful and it can occur at other times of the day besides early morning. It occurs when the patient is off, and the plasma levels of dopa are low. However, it is not just the low plasma levels, because if you give the patients no more drugs, in a couple of days they become very parkinsonian, and eventually that dystonia will disappear. It is not part of the disease. That type of off-dystonias versus the kind you just talked about, Roger, are quite different. Then, of course, there are the peak-dose dystonias and the diphasic dystonias and so forth. I think they are all different, with different mechanisms.

Yébenes: Coming back to Dr. Melamed's question. I think, there is increasing evidence that the mechanism explaining dystonia in parkinsonism is disequilibrium between high norepinephrine transmission and low dopamine transmission. I am going to try to summarize the evidence so far available. First of all, F. Bermejo did a study, that I am coauthor with, showing that off-dystonia in Parkinson's disease, either early-morning dystonia or off-period dystonia, disappears in a few days after the patients are taken off from dopa. When dopa is started again after at least one week of being off, dystonia reappears, but not at the first doses; it takes a few days until dystonia appears again. Victoria Muradas did a study in my group showing that when you treat normal rats with chronic levodopa, with high doses of carbidopa, norepinephrine metabolism is increased but does not immediately increase. MHPG-levels after one dose of L-dopa do not change into the brain, but after one week of treatment MHPG-levels into the brain go to 300% of normal values. This means that chronic treatment with L-dopa could increase norepinephrine metabolism, but it does not happen immediately. Evidence that norepinephrine versus dopamine disequilibrium could be responsible for dystonias is provided by some neurochemical studies performed in brain.

Psychiatric problems in treating Parkinson's disease

Duvoisin: I can remember the hallucinatory states and confusions that we saw with anticholinergic drugs prior to the use of levodopa. My judgement is that there was much more cytotoxicity with the common anticholinergics than we see with levodopa. I wonder if you'd comment on that.

Klawans: I think that the therapeutic index of anticholinergics is terrible and the older the patients get, the worse the psychotoxicity. The presence of any degree of dementia, I think, puts the therapeutic index of anticholinergics well below one. On the other hand, I think that if you use enough anticholinergics you can make anybody hallucinate whether or not their parkinsonism got better.

Marsden: I wonder if we could have some discussion on the duration of the psychosis, and on the dissociation between the psychosis and the motor effects of dopaminergic stimulation. I am sure all of us appreciate that the persistence psychosis nearly always exceeds the duration of motor action of levodopa. You stop levodopa, and the psychosis goes on for many days if not weeks. However, I have never looked carefully to see whether the duration of the psychosis parallels the very long-term motor effects of levodopa. Stop levodopa and patients revert back to the motor state they are on in their usual off-periods. Two or three or four days later they get disastrously worse. This is the drug holiday experience that most of us have. Does the duration of the psychosis parallel the delayed motor deterioration that occurs on stopping drug treatment?

Klawans: It more tends to follow the delayed motor performance. In our experience, patients with a significant paranoid psychosis, in whom the drug is withdrawn, get better the next day, or two or three days later.

Stahl: You have shown peak-dose psychoses. Have you ever seen, or is there any discussion in this room, about off-period depression, particularly mood swings, and the evolving of a kindled syndrome where the off-period is a positive phenomenon? We have withdrawal depression alternating with euphoria or at least euthymia.

Duvoisin: I'd like to ask you, have you seen depression, simple depression, as a complication of levodopa therapy?

Klawans: That is even harder to answer than the questions in relation to the psychoses and/or hallucinations. Depression is common enough in the control population at any time in the United States. It certainly is common in anyone with progressive severe disability from a wide variety of diseases. It is certainly commoner in Parkinson's disease than it is in the control population. It does not, as far as I can tell, in our studies actually parallel the severity of disability in Parkinson's disease. Pathophysiologically related is the group that Steve Stahl has just referred to, namely those patients who have depression of very short term related to their off-periods. I would think that whatever the mechanism of "off" is, and perhaps one of the things that would make one think that "off" may be related to non-dopaminergic factors, is that percentage of patients who become depressed when off. I think it is rare, I haven't heard of, to be manic when on.

Quinn: In the last couple of years we have seen three patients who have had off-period psychoses, not just off-period depression. And this got me wondering about what we always do in psychotic patients. If the patient is hallucinated or paranoid or deluded late at night, we say it must be too much dopa. We therefore stop their late levodopa doses. I just wonder whether amongst this group there are a number of people who are like that because we are not giving them enough, rather than too much.

Klawans: That may be possible, but I don't think I have guts enough when the patient is already psychotic to add more medication. On the other hand, I think that patients certainly can become psychotic during off-periods. What I forgot to mention is that not only are there patients whose psychosis persists longer than their motor benefit, but their psychosis worsens during drug withdrawal. On the 3rd–7th day off they are clearly more psychotic, more out of touch with reality, and may first become confusional.

Marsden: Concerning the problem of sustained depressive mood in Parkinson's disease, Brown and Gothan in our department compared a group of patients with Parkinson's disease with an equally disabled group of patients with rheumatoid arthritis. They were unable to demonstrate any difference in the prevalence of depression in the two groups. My memory is that untreated patients with desperately severe Parkinson's disease before levodopa was invented occasionally could have the hallucinosis in clear sensorium of the sort that we are describing here. This was a predominantly visual non-threatening hallucinosis with an absolutely clear sensorium. Would the elder statesmen confirm or refute that that can be part of the untreated Parkinson's disease?

Duvoisin: I'll confirm it and we still see it.

Klawans: I also think we still see it. The real question is, what percentage of those people in whom such a hallucinosis persists of medicine would have ended up with that hallucinosis anyway as part of their progressive not purely nigral disease. It may well be a peduncular hallucinosis, a midbrain hallucinosis, whatever that is. There are many characteristics that are very similar to those descriptions. As a matter of fact, we have within the last few years in patients with progressive continuous hallucinosis put them back on the drug. If they are going to hallucinate they might as well walk while they are talking to their "friends" as not.

Marsden: Can those who have heard Bill Langston describe the MPTP-patients, who are reputed not to have mesolimbic dopamine deficiency, tell us what the psychiatric phenomena in those patients, untreated, is and what the effects of levodopa is. It has not come over clearly to me in the publications what psychiatric problems those patients have developed.

Fahn: I can't answer it from personal observations myself. I have never seen one of those patients with parkinsonism from MPTP. But I do recall Bill mentioning patients having mental problems. I don't think he really deciphered the psychiatric aspects in detail. Don't forget these were drug addicts, and they had some kind of personality changes already. So it wasn't clear that their problems arose from the disease or the loss of striatal dopamine.

Gessa: Dopamine seems to be the common mediator for drug abuse. I am asking whether dopa is abused or lisuride is abused, and if not why it is not abused?

Klawans: There are patients who clearly do abuse. One of my patients tends to stay manic all the time with severe dyskinesias, I think he sees three or four different neurologists and keeps collecting more and more prescriptions, basically for that reason. I know of no abuse of it on the marketplace. None of the dopa-abusers began abusing dopa within the first two days of taking it. I think also it is not an easy drug to take. Most normals would not feel terribly good taking it.

To go back to the previous question, I do have one MPTP-induced Parkinson patient in our population. He is secretive and paranoid, and I think he is a bit more secretive and more paranoid than he was prior to be on treatment, but that is hard to prove with hard data.

Duvoisin: That the pathological substrate will affect the manifestations would be very clear from the old experience of postencephalitics. You may remember Oliver Sacks', "Awakenings" on the response of a sanatorium full of post-encephalitics to levodopa treatment. He certainly saw their compulsive motor behaviours and rituals and psychopathic behaviour recur, phenomena that we did not see in Parkinson's disease patients.

Dorow: It is just a comment on the effects of dopamine agonists on sleep. We have tested lisuride and other dopamine agonists in healthy subjects. In the first half of the night when lisuride is given intravenously we see a clear suppression of REM sleep. In the second half of the night there is a rebound REM, which corresponds nicely to the clearance of the drug. Bromocriptine, by the way, has the same effect and also pretreatment with sulpiride inhibits this effect of the dopamine agonists. In healthy subjects we do see this REM suppression. Maybe the vivid dreaming you can see in some of the patients may be due to a REM rebound.

Klawans: Sleep within a Parkinson population, is a very difficult problem. Untreated Parkinsons' have decreased REM sleep. On levodopa therapy there is increased REM. It is late in dopa therapy that the REM tends to be suppressed. The real question in relation to these hallucinations is whether they are not merely the leaking over in wakeful consciousness of dream material. Many of the patients who say they are hallucinating during the day, when we have then done continuous EEGs actually had very brief naps, NREM dreams. They awoke and said they had hallucinated. So some of these patients actually report as a hallucination what really may have been a dream state.

Tolosa: The early literature on levodopa included depression as a common side effect. Do you think that levodopa induces depression? I might say that we have seen several patients that want to commit suicide soon after they take levodopa.

Klawans: I am not terribly convinced, other than by the off-associated depression. In over 3000 patients I have treated I have had only one suicide in 18 years, which I think is less than the control population in the United States.

Parkes: Perhaps we should stress what we don't see, which is, of course, schizophrenia. And the second thing we don't see, Prof. Stepherd at the Institute of Psychiatry showed me a patient last week with a missing link, a combination of night terrors and waking panic, agoraphobic attacks. And of course, night terrors are familial and presumably biochemical. And I am not sure that I have seen waking panic or agoraphobia in any levodopa-treated parkinsonian. I am not entirely sure whether you are relating sleep problems with subsequent development of psychosis.

Klawans: To deal with your first question, I think I have seen more panic related to off than I have depression. I have patients who each time they have an off-period have an absolute panic attack. I think their panic attacks are common.

Parkes: On-period panic attacks?

Klawans: I think I have seen on-period, too, but predominantly off-period panic attacks. Secondly, I think sleep disruption is a predictor of dopa-induced psychotic disease, but has nothing to do with sleep apnoea. In our patient population, which is the only study that I know of that has been done, paranoid psychosis does not occur among those people without sleep disruption.

Toone: Just to respond to D. Parkes' point. I agree that one doesn't see psychotic reactions which take on the course of schizophrenia, but you see psychotic reactions which as they present are schizophrenic in form. I think when you are describing, I think earlier people we have heard, paranoid delusions, these are usually indistinguishable from schizophrenia at the point of examining the mental states. But I agree, they were off within a few days of drug withdrawal.

Marsden: Just a very quick comment which has a practical implication concerning suicide in Parkinson's disease. One of the striking differences between the usual depression of Parkinson's disease and primary depression is the lack of self-blame and guilt.

Klawans: The primary feature in the depression of the Parkinson patients is what my neuropsychologists now call an anhedonia as opposed to a guilt-ridden phenomenon.

Continuous dopaminergic stimulation experimental studies

J Neural Transm (1988) [Suppl] 27: 141–160
© by Springer-Verlag 1988

Continuous intracerebroventricular infusion of dopamine and dopamine agonists through a totally implanted drug delivery system in animal models of Parkinson's disease

J. G. de Yebenes[1], S. Fahn[2], V. Jackson-Lewis[2], P. Jorge[1], M. A. Mena[1], and J. Reiriz[1]

[1] Centro Ramon y Cajal, Madrid, Spain
[2] Columbia University, New York, New York, U.S.A.

Summary. We studied the effect of intracerebroventricular infusion of dopamine and dopamine agonists in animal models of dopamine deficiency as an experimental approach to the treatment of levodopa induced fluctuations in Parkinson's disease. Dopamine deficiency was produced in rats by unilateral lesion of the nigrostriatal pathway or by chronic treatment with reserpine. Monkeys were lesioned by intravenous injection of MPTP. The animals were treated with intracerebral infusions of dopamine (with or without associated intraperitoneal administration or intracerebroventricular infusion of pargyline), lisuride and pergolide. The intracerebroventricular infusion of these drugs was performed with osmotic minipumps in rats and with infusaid pumps in the monkeys. The infusion of dopamine or dopamine agonists in rats with unilateral lesions by 6-OH-dopamine produced a persistent rotation contralateral to the lesioned and implanted side. The infusion of dopamine reversed reserpine-induced akinesia only when pargyline was associated. In the range of concentration used, maximum allowed by solubility of compounds, the effects of dopamine were more potent than those of the agonists. In spite of the stability of dopamine "in vitro" when dissolved in antioxidants and at low pH, a pigment, product of autooxidation, was found in the brains of the animals infused with dopamine. The monkeys were implanted with infusaid pumps and infused for up to 3 weeks. The pump was not well tolerated due to its huge size for the animals. One monkey showed reversal of the MPTP-induced akinesia while the other, whose catheter had moved from the correct implantation site, remained unchanged. In both monkeys there was evidence of autooxidation of dopamine. Intracerebral infusion of dopamine agonists may be a possible experimental alternative to the treatment of levodopa induced fluctuations in Parkinson's disease but stable and soluble dopamine agonists and suitable delivery systems are needed.

Introduction

Levodopa is the most effective drug for Parkinson's disease (PD) but in the course of treatment many patients develop adverse effects, including treatment-related motor fluctuations, dyskinesias and psychiatric complications (Marsden and Parkes, 1977; Fahn and Calne, 1978; Lesser et al., 1979; Rinne, 1981; Marsden et al., 1982). Several lines of evidence suggest that some of the adverse effects are at least partially related to pharmacokinetic mechanisms. Clinical and experimental studies have shown that: a) there is good correlation between motor fluctuations and abnormalities of absorption and/or peripheral metabolism of DOPA in patients with PD (Nutt et al., 1984; Mena et al., 1986); b) intravenous levodopa infusions improve or supress motor fluctuations (Quinn et al., 1982; Quinn et al., 1984; Marion et al., 1986); c) the administration of drugs or foods containing amino acids that interfere with levodopa absorption or transport to the brain block the beneficial effect of levodopa infusion (Nutt et al., 1986).

Levodopa infusions are valuable in understanding the pathophysiology of abnormal responses but are of little practical use due to the poor solubility of levodopa in water at neutral pH and to the need of administering it in large volumes of fluid (Quinn et al., 1982). This limitation could be overcome by the use of soluble precursors of levodopa or water soluble dopamine agonists. Levodopa methyl ester is a precursor of levodopa that is effective in reversing reserpine-induced akinesia in rats. Its chronic administration to humans may be unsafe, however, since levodopa methyl ester is metabolized to methanol in the body (Juncos et al., 1986).

Lisuride, a water soluble DA agonist, has been administered as a continuous subcutaneous infusion through an external pump in more than 40 patients (Obeso et al., 1986). In most of these patients lisuride infusions improved motor fluctuations. This therapeutic approach has, however, limitations including day-to-day variability in clinical performance due to difference in absorption, psychotic reactions, and production of skin nodules and skin reactions in some patients. Moreover, most patients could not be controlled by lisuride alone and needed the concomitant administration of oral levodopa. Psychiatric complications, including acute toxic psychosis, required discontinuation of therapy in one-third of the patients. Hypersexuality was present in a smaller percentage.

Intracerebroventricular infusion (IVC) of DA or DA agonists through a totally implanted drug delivery system may have some theoretical advantages over peripheral infusion. It may be possible to deliver the endogenous neurotransmitter, rather than precursors requiring metabolism or synthetic DA agonist with non-selective pharmacological profile of activity. In addition, pharmacokinetic abnormalities of absorption and transport could be totally avoided by delivering the drug at its site of action. Finally, it may be possible to prevent the appearance of central side effects, presumably caused by stimulation of DA receptors in brain areas other than the nigro-striatal system (the mesolimbic system is presumably the system in which DA produces psychiatric symptoms and the hypothalamic DA system is the site where DA induces hypersexuality) by directly delivering the active drug to the target brain region, the striatum.

In order to test the feasibility of intracerebral infusions of DA or DA agonists in patients with PD we performed the following experiments in animal models of PD.

Procedures

Stability of DA "in vitro"

DA stability was evaluated in light-protected solutions maintained at $37\,^\circ$C up to one week. Stability was calculated by measuring DA concentrations at different time intervals and by measuring the concentration of its auto-oxidized metabolites. DA concentration was measured by high performance liquid chromatography with electrochemical detection (HPLC/ED). Autooxidation of DA was evaluated by spectrophotometry measuring enhancement in absorbance at $480\,$nm caused by the presence of autooxidized derivatives of DA (Van der Werde, 1964). DA was dissolved in different vehicles including double distilled water (DDW), 5% absorbic acid in DDW, 0.5% $Na_2S_2O_5$ and artificial CSF.

Experiments with rodents

Different models of DA deficiency were produced in male, Sprague-Dawley albino rats, weighing 220–250 grams at the beginning of the experiment. Intracerebroventricular (ICV) infusions of DA and DA agonists were tested to reverse the behavioral and biochemical abnormalities seen in these animal models. All animals were kept in plastic cages, 4 to 6 animals per cage. The animals consumed regular lab chow and water "ad libitum" and were maintained at $22\,^\circ$C, with an artificial 12 hour interval light and darkness cycle.

Implantation of the animals

Rats were implanted into the left cerebral ventricle with a cannula connected to an Alzet osmotic minipump, subcutaneously placed in the interscapular region. Cannulae were prepared from polyethylene tubing external diameter $0.965\,$mm. This tubing was cut in pieces and at $4\,$mm from the proximal end a small dilatation was made by placing this section of the tubing in a heating system made from a circular resistence connected to a 9 volt battery. This dilatation would prevent the cannula from moving since it would be held by the skull below it and the cement above it. Pumps were filled with vehicle ($0.01\,$N HCl or 0.5% $Na_2S_2O_5$) or with different concentrations of DA, pargyline (P), deprenyl (D), DA + P, DA + D, lisuride ($1\,$mg/ml), or pergolide ($1\,$mg/ml). The rats were anesthetized with sodium pentobarbital, $7\,$mg/kg ip, and placed into a stereotactic apparatus with the nose bar $5\,$mm above the horizontal plane. A midline longitudinal incision was made in the scalp and the bregma was explored and measured. At the point of insertion of the cannula (+ 0.08 AP, + 0.13 Lat) a hole was drilled and the dura mater underneath was punctured. Two additional holes were drilled in nearby places and two 3 mm jewelry screws were implanted in them to secure the immobility of the cannula. The proximal tip of the cannula was inserted into the left ventricle (4 mm DV in the hole previously drilled) and the distal tip was dilated with a pair of fine forceps and connected to the flow restrictor of a previously filled osmotic minipump. The cannula was then cemented in place with dental cement and the pump was pushed through a previously prepared subcutaneous tunnel made at the caudal end of the surgical incision in the interscapular region.

Biochemical analysis of brain regions

Animals were killed by decapitation and brains were quickly removed. Nucleus accumbens, dorsal striatum, cerebral cortex and hippocampus from both cerebral hemispheres, as well as hypothalamus, brain stem and cerebellum were dissected, collected separately, frozen on dry ice, wrapped in aluminum foil, and kept frozen at $-80°$ until assayed for catecholamines, DOPAC, and DA receptors. Catecholamines and DOPAC were measured by HPLC/ED previously described techniques (Reches and Fahn, 1982). DA receptors were measured by the specific binding of ^{3}H-spiperone to striatal membranes according to Seeman et al. (1975) and Muradas et al. (1986).

ICV infusion of DA in catecholamine depleted animals by α-methyl-para-tyrosine (AMPT)

Animals implanted with cannulae connected to pumps filled with DA or vehicle, were treated on day 6 of infusion with AMPT, 250 mg/kg, 90 minutes before death. Motor activity was evaluated in a Columbus Instruments activity meter (Columbus, OH) by accumulative counts recorded during a 30 minute interval from 35 to 5 minutes before death. Animals were sacrificed by decapitation and catecholamines measured in different brain areas according to the above described techniques.

ICV infusion of DA and DA agonists into rats with unilateral lesions of the nigro striatal pathway by 6-OH-DA

The left nigrostriatal DA pathway was lesioned by stereotactic injection of 6-OH-DA (8 µg in 4 µl of 0.2% ascorbic acid) into the anterior substantia nigra, as previously described (Reches et al., 1982). Three to five weeks after the denervation the animals were treated with apomorphine 1 mg/kg s.c. The animals making more than 30 turns contralateral to the lesioned side per 30 minutes period were considered well-denervated, and were matched and assigned to different experimental and control groups with comparable degrees of contralateral rotation. The animals were implanted into the left cerebral ventricle as previously described, with a catheter connected to an Alzet 2001 osmotic mini-pump filled with DA, DA + P, lisuride, pergolide or vehicle (0.5% $Na_2S_2O_5$). "Spontaneous" rotation was measured daily during the ICV infusion and apomorphine-induced rotation was measured the first and the last day of infusion. The animals were killed 6 days after implantation, their brain dissected and catecholamines, DOPAC and DA receptors measured in brain regions.

ICV infusion of DA and dopamine agonists in reserpinized rats

Animals treated with unilateral ICV infusion of DA, DA + P, DA + D, pergolide or lisuride were treated with reserpine according to the treatment schedules to be described below. Motor activity was measured daily for periods of 10 minutes on a Columbus Instruments activity meter, or for periods of 30 minutes in activity meters designed and manufacturerd by Dr. Parrẽno, C. Ramon y Cajal (Madrid, Spain). The animals were killed 6 days after implantation and their brains dissected out and analyzed for catecholamines, DOPAC and DA receptors as previously described.

The effect of ICV infusion of DA on reserpine-induced akinesia was tested in different experimental situations.

— ICV infusion of DA through external pump
Animals with cannulae previously implanted into the left ventricle were treated with a single dose of reserpine 10 mg/kg s.c. When akinesia was severe, 16–20 hours later, DA 15–20 μg, was infused at a rate of 1 μg/min for a period of 15–20 minutes.

— Intracerebroventricular infusion of DA through a totally implanted drug delivery system

Experiment A
Rats were implanted with a cannula connected to a pump filled with DA (100 mg/ml in 0.5% $Na_2S_2O_5$) or vehicle. Then the animals were treated with reserpine s.c. 0.5 mg/kg/day for 6 days. Motor activity was measured daily 16–20 hours after the last reserpine dose. Animals were killed 6 days after the implantation and dissected brain regions were used for the analysis of catecholamines and DOPAC.

Experiment B
In addition to the protocol described in Experiment A, all rats were treated with pargyline 25 mg/kg ip daily, 16 hours after every reserpine injection. Motor activity was measured in all animals for 10 minutes every hour for 6 hours after pargyline.

Experiment C
Animals were treated as in Experiment A. The control group received ICV infusion with vehicle: the experimental groups were infused with pargyline (25 mg/ml) and DA (5–100 mg/ml) + pargyline (25 mg/ml). Similar experiments

were performed in animals receiving identical treatment but in which the pump was filled with deprenyl (10 mg/ml) instead of pargyline, or with lisuride (1 mg/ml), or pergolide (1 mg/ml) in place of DA.

Experiments with monkeys

Two cynomolgus monkeys (*Macaca fasciculata*) were implanted with chronic delivery systems (Infusaid pumps, Infusaid, Milford, MA). One animal served as a control: the other was made severely akinetic by MPTP treatment (2 mg kg i.v., in 4 divided doses of 0.5 mg/kg i.v. every 2 hours). One week later, when it was akinetic, the ICV infusion with vehicle was replaced by another containing DA (25 mg/ml) + Deprenyl (2 mg/ml). After 42 days of total infusion and 9 days of infusion with DA + Deprenyl the monkeys were killed by pentothal anesthesia, and catecholamine and DOPAC were analysed in brain areas.

Statistical analysis

Analysis of variance was performed followed by student's t test or Wilcoxon's test.

Results and discussion

Solubility and stability of DA and DA agonists in different vehicles

DA was soluble in water at least up to concentrations of 200 µg/ml. At neutral pH and 37 °C DA rapidly auto-oxidized even in the dark. The rate of auto-oxidation was three times higher in the CSF than in double distilled water (Fig. 1). Ascorbic acid (0.2–5%) and EDTA (10%) did not prevent DA auto-oxidation, but sodium metabisulfite (0.05–0.5%) greatly reduced it. The concentration of DA in 10^{-2} M HCl or 0.5% $Na_2S_2O_5$ at 37 °C remained stable for one week in the dark.

ICV infusion of DA in AMPT treated rats

AMPT treatment, as expected, reduced DA and DOPAC concentrations in the striatum to 75% of baseline levels 90 minutes after treatment (Fig. 2). ICV infusion of DA (0.5–100 mg/ml) did not reverse but even enhanced AMPT-induced depletion of DA (Fig. 2). AMPT also reduced motor activity, and this reduction was not counteracted by DA infusion.

The effects of ICV infusion of DA on motor activity and AMPT-induced depletion of monoamines were unexpected. Since identical results were obtained in animals infused with both low (0.46 µg/h) and high (92 µg/h) dose of DA, it is unlikely that the reduction in motor activity and DA concentrations are caused by the selective action of DA on DA autoreceptors. It is likely, however, that chronic infusion of DA for 5 days in normal rats down regulates DA receptors or reduces the activity of uptake sites to an extent that the DA cell is unable to compensate by the blockade of catecholamine synthesis in the short period of 90 minutes.

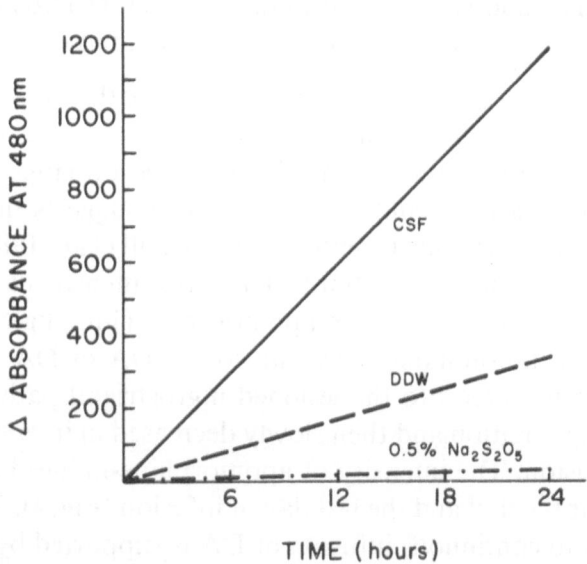

Fig. 1. Auto-oxidation of DA "in vitro" in different media. Aminochrome formation was measured by the increment in optical density at 480 nm. Values represent the mean of three samples

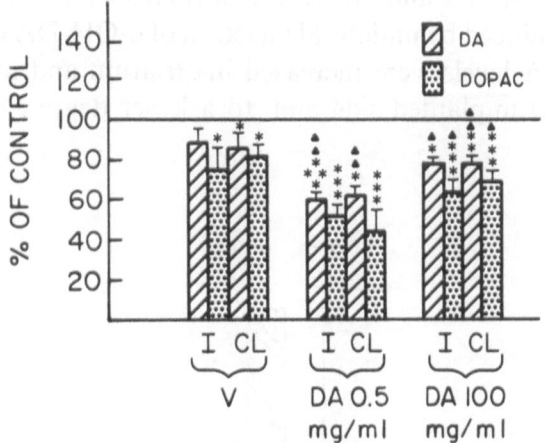

Fig. 2. Effect of ICV infusion of DA on AMPT-induced depletion of catecholamines. Values represent the mean ± S.E.M. of 8 animals in percentage of control levels. * $p < 0.05$, ** $p < 0.01$, *** $p < 0.001$ respect to the control animals, ▲ $p < 0.05$, ▲▲ $p < 0.01$ respect to vehicle-infused AMPT-treated animals. *I* Infused side, *CL* contralateral side. *V* AMPT treated vehicle infused animals. *DA 0.5* AMPT-treated, infused with DA 0.5 mg/ml; *DA 100* AMPT-treated, infused with DA 100 mg/ml

ICV infusion of DA and DA agonists in rats with 6-OH-DA-induced unilateral lesions of the nigro-striatal pathway

Rats with unilateral lesions of the nigrostriatal pathway, implanted with a catheter connected to a pump delivering DA or DA agonists into the ventricle ipsilateral to the lesion, showed contralateral rotation (Figs. 3 and 4) during the infusion of DA, lisuride and pergolide, which suggests that in this model these drugs act at postsynaptic receptor sites (Reavill et al., 1983). The intensity of the rotation at peak activity is about 10% of the intensity of rotation induced in the same animals by high dose of apomorphine (1 mg/kg).

Contralateral rotation induced by infusion of DA or DA agonists into the cerebral ventricle ipsilateral to the lesioned nigrostriatal pathway peaked two days after the implantation and then slowly decreased in intensity over a period of 5 days of infusion. The intensity of apomorphine-induced contralateral rotation was similar the first and the last day of infusion (Fig. 4). The development of tachyphylaxis to continuous infusion of DA is supported by the finding that DA receptor binding of spiperone, increased on the vehicle-infused lesioned side, returned toward normal values after two days of infusion and to values lower than the intact side after 7 days of infusion (Fig. 5). Scatchard analysis of the binding data revealed no evidence of changes in affinity in DA and lisuride infused animals.

DA infusions restored and surpassed normal DA tissue levels, whose depletion had been induced by unilateral injection of 6-OH-DA into the substantia nigra (Table 1). DA levels were increased in striatum and nucleus accumbens in the lesioned and implanted side and, to a lesser degree, in the intact, con-

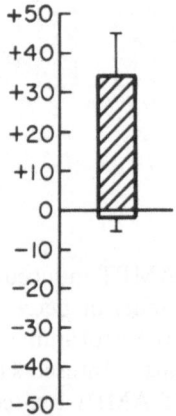

Fig. 3. "Spontaneous" rotation in 6-OH-DA lesioned, vehicle and DA-infused rats. Animals with unilateral 6-OH-DA lesion of nigrostriatal pathways, responding with contralateral rotation to apomorphine, 1 mg/kg, s.c., were infused with DA (100 mg/ml) or vehicle (0.5% $Na_2S_2O_5$), into the ventricle ipsilateral to the lesioned side. Turning behavior was measured at peak of activity between 24 and 48 h. Contralateral rotation to the lesioned and implanted side is represented as + and ipsilateral as −

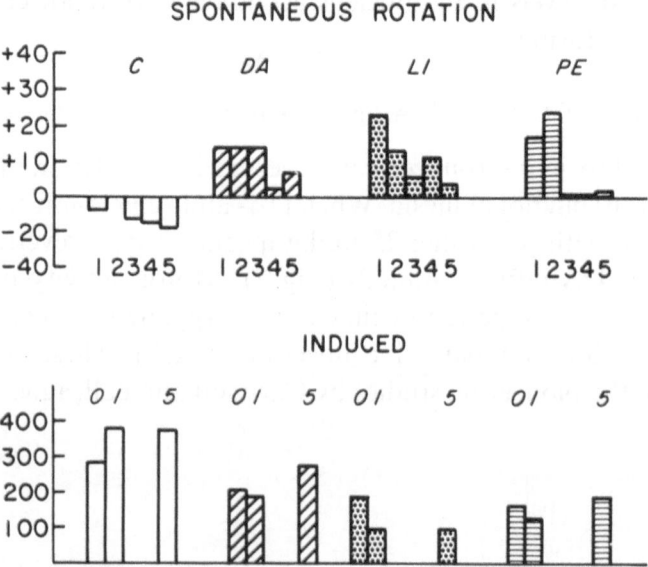

SPONTANEOUS ROTATION

INDUCED

Fig. 4. Spontaneous and apomorphine-induced rotation in 6-OH-DA (100 mg/ml), lisuride (1 mg/ml) and pergolide (1 mg/ml). Spontaneous rotation was measured for 30 minutes every day during the period ICV infusion. Apomorphine-induced rotation was measured for 30 minutes, after treatment with apomorphine 1 mg/kg s.c., on days 1 and 5 of the ICV infusion period and previous to the implantation of the animals. Positive rotation means contralateral to the lesioned and implanted side, negative means ipsilateral

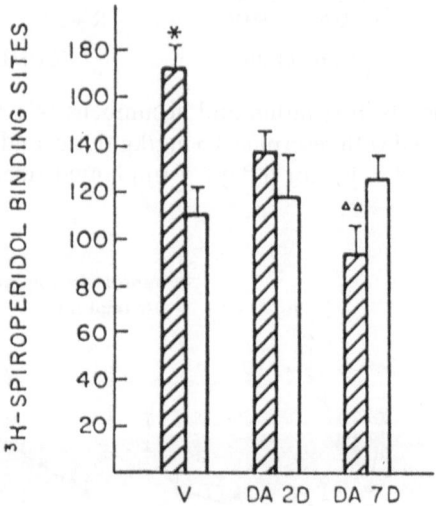

Fig. 5. ^{3}H-Spiperone binding in 6-OH-DA-lesioned DA-infused rats. Values represent the mean ± S.E.M. (6 animals) of specific ^{3}H-spiperone binding to striatal membranes. Striped bars represent binding to the denervated and infused striatum, while empty bars represent binding to the contralateral side. * $p < 0.05$ respect to the contralateral side of the same group of animals. $\triangle \triangle$ $p < 0.01$ respect to the same side of vehicle implanted animals. V Vehicle, *DA2D* 2 day after infusion of dopamine, *DA7D* 7 days after infusion of dopamine

tralateral side. NE levels in the nucleus accumbens were not changed by the lesion or by the infusion.

ICV infusion of DA and DA agonists in reserpine-induced akinesia

ICV infusion of DA in reserpinized rats reversed DA depletion in striatum and accumbens (Fig. 6) but not akinesia. When DA- and vehicle-infused reserpinized rats were treated with pargyline, 25 mg/kg/ip once a day, reversal of akinesia was observed in DA-infused animals (Fig. 7). Motor activity of DA-infused animals increased over a period of time after pargyline treatment and reached its peak at 6 hours after pargyline injection (Fig. 8). These findings are in agreement with the pioneering studies by Carlsson and colleagues showing that

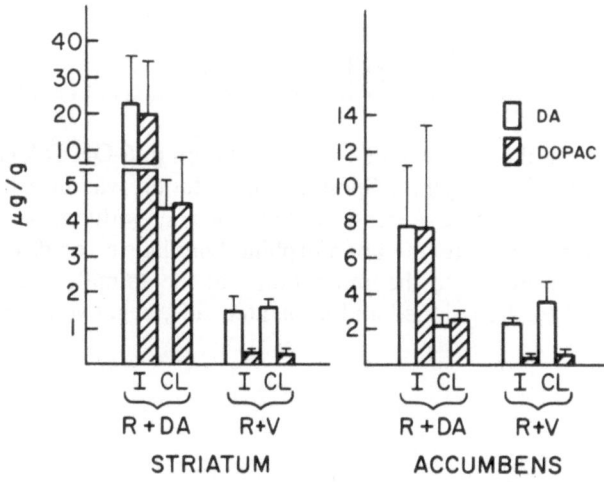

Fig. 6. DA and DOPAC levels in striatum and accumbens of reserpinized DA-implanted rats. The animals were treated with reserpine 0.5 mg/kg/d. s.c. and infused with DA 100 mg/ml ($R + DA$) or vehicle ($R + V$), for 6 days. I Implanted side, CL contralateral side

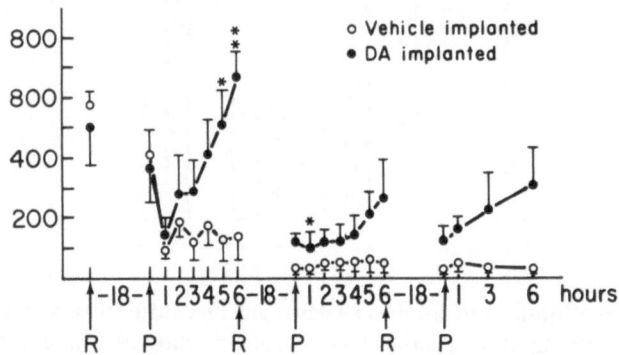

Fig. 7. Motor activity in reserpinized rats during ICV infusion with DA (100 mg/ml) or vehicle, and parenteral treatment with pargyline, 25 mg/kg/d. Pargyline was given i.p. 18 h after reserpine. Motor activity was measured at intervals of 10 minutes before and every hour after pargyline injection. * $p < 0.05$, ** $p < 0.01$ respect to vehicle implanted group

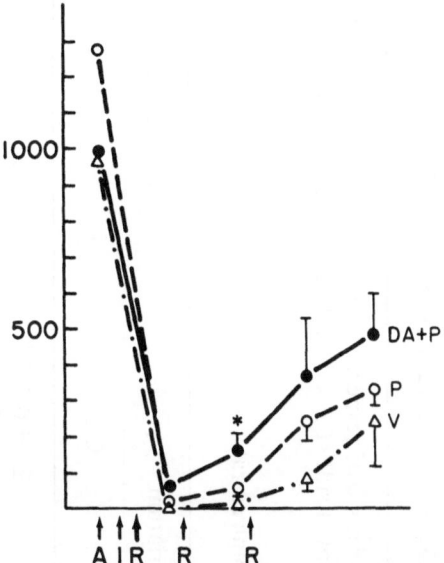

Fig. 8. Motor activity in reserpinized rats during intracerebral infusion of vehicle (*V*), pargyline (*P*), 25 mg/ml, and DA (100 mg/ml) with P (25 mg/ml) (*DA + P*). Baseline motor activity (*A*) was measured before the implantation (*I*) of a catheter. Animals were treated with an initial dose of reserpine (*R*) 1 mg/kg s.c., followed by 0.5 mg/kg/d. * p < 0.05 respect to V group

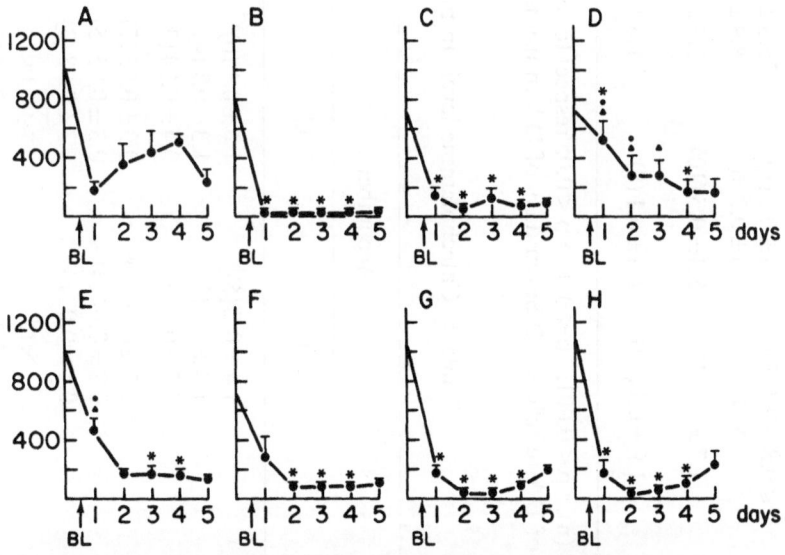

Fig. 9. Motor activity in controls (**A**) and reserpinized rats after cerebral infusion of vehicle (**B**), pargyline (P) 25 mg/ml (**C**), DA, 100 mg/ml, + P (**D**), DA, 25 mg/ml, + P (**E**), DA, 5 mg/ml, + P (**F**), Lisuride, 1 mg/ml (**G**), and Pergolide, 1 mg/ml (**H**). * p < 0.05 respect to the control, group A. ▲ p < 0.05 respect to reserpinized, vehicle infused animals, group B. ● p < 0.05 respect to reserpinized, pargyline infused animals, group C. Baseline motor activity (*BL*)

Table 1. Catecholamines and DOPAC in corpora striata and accumbens of 6-OH-DA-lesioned, DA-infused rats

		C		2D		7D	
		LI	CL	LI	CL	LI	CL
Striata	DA	≤0.8±0.1(8)[a]	8.3±0.3(8)	26.8±10.8(7)	24.4±15.17	37.5±14.9(5)[b]	11.3±7.0(70)
	DOPAC	ND	1.6±0.2(6)	8.9±3.3(6)	4.0±1.1(6)	8.1±3(10)[b]	3.1±0.9(70)
Accumbens	DA	1.7±0.2(10)[b]	5.4±0.5(10)	2.5±0.5(5)[c]	5.6±1.0(7)	8.8±2.8(9)[d]	7.2±1.4(4)
	NE	1.9±0.1(10)	1.8±0.1(8)	1.6±0.3(7)	1.6±0.3(7)	2.4±0.2(8)[c]	2.2±0.3(9)

Values express X ± S.E.M. (n). [a] $p < 0.001$, [b] $p < 0.01$, [c] $p < 0.05$ respect to the control, intact side. [d] $p < 0.05$ respect to vehicle infused, lesioned side. C Control. 2D Two days of DA infusion. 7D Seven days of DA infusion. LI Lesioned, implanted side. CL Contralateral side. ND Not done

Table 2. Catecholamine levels in brain regions of reserpinized rats

		Accumbens		Striatum		Hypothalamus
		I	CL	I	CL	CL
DA	DA+P	1.4±0.2(6)[b]	1.6±0.1(6)[a]	2.2±0.4(6)[c]	2.1±0.2(7)[b]	0.5±0.1(5)
	P	1.4±0.1(3)[a]	1.3±0.1(4)[a]	1.3±0.2(5)[c]	1.4±0.2(5)[c]	0.3±0.03(5)
	C	0.5±0.1(5)	0.4±0.04(3)	0.6±0.03(4)	0.7±0.1(4)	0.3±0.1(4)
NE	DA+P	0.3±0.06(7)[c]	0.3±0.05(7)	0.1±0.01(5)[d]	0.1±0.03(6)	0.4±0.07(7)
	P	0.2±0.04(4)[c]	0.3±0.05(5)	0.2±0.02(4)	0.2±0.01(5)	0.4±0.07(5)
	C	0.1±0.02(4)	0.3±0.08(3)	ND	ND	0.2±0.06(4)
DOPAC	DA+P	ND	0.3±0.09(2)	0.3±0.1(3)[a]	0.3±0.06(5)[a]	
	P	ND	0.3±0.05(2)	0.2±0.02(2)[a]	0.3±0.03(2)[a]	
	C	0.6±0.04(5)	0.5±0.05(3)	1.0±0.02(5)	0.9±0.04(5)	

Values express X ± S.E.M. (n). [a] $p < 0.001$, [b] $p < 0.01$, [c] $p < 0.01$ respect to the same side of controls (reserpine-treated, vehicle infused animals). [d] $p < 0.01$ respect to the same side of Pargyline infused animals. I Implanted side, CL contralateral side. DA + P Reserpinized rats infused with DA (100 mg/ml) + Pargyline (25 mg/ml), 0.92 ml/h; P reserpinized rats infused with Pargyline (25 mg/ml), 0.92 ml/h; C reserpinized rats infused

levodopa reverses reserpine-induced akinesia in iproniazide-treated mice (Carlsson, 1959; Carlsson et al., 1957). Pharmacological differences between early and advanced Parkinson's disease may be compared to the animal models of catecholamine depletion by AMPT and monoamine depletion by reserpine. It is interesting to note that in our animal model where synaptic storage of the neurotransmitter is impaired by reserpine, it is impossible to reverse behavioral changes unless one uses MAO inhibitors in addition to DA. That may raise important implications for the treatment of advanced PD (Marsden, 1980).

When reserpinized rats were implanted with cannulae connected to pumps delivering vehicle, pargyline, or DA + pargyline it was found that reversal of akinesia was quicker in animals implanted with DA + pargyline than in the other two groups (Fig. 8), but the differences only reached significant values 48 hours after implantation. In striata ipsilateral to the cannula DA levels increased 2-fold in pargyline-implanted animals and 4-fold in DA + pargyline-implanted animals and 2 to 3 fold in the contralateral striata and nuclei accumbens (Table 2). DOPAC levels were decreased in both ipsi- and contralateral striata and accumbens in pargyline and DA + pargyline-treated rats, though DOPAC reduction was greater in the ipsilateral side. NE levels were increased also in both striata and nuclei accumbens. Reversal of reserpine-induced akinesia in rats by ICV infusion of DA + pargyline was dose-dependent (Fig. 9). Lisuride and pergolide, at maximal concentrations allowed by its solubility in water (1 mg/ml), did not have any anti-akinetic effect during the first 3 days of infusion, but they improved motor activity during the last two days of the experiment. This phenomenon, increasing potency with time, has been observed by others

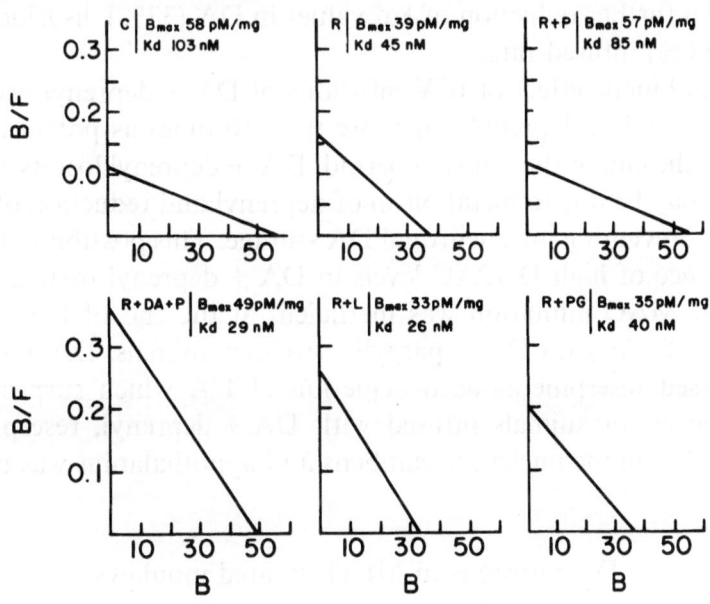

Fig. 10. Scatchard plots of ^{3}H-spiperone binding to striatal membranes of controls (*C*) and reserpinized animals infused with vehicle (*R*), pargyline, 25 mg/ml, (*R* + *P*), DA, 100 mg/ml + P (*R* + *DA* + *P*), Lisuride, 1 mg/ml (*R* + *L*), and Pergolide, 1 mg/ml (*R* + *PG*)

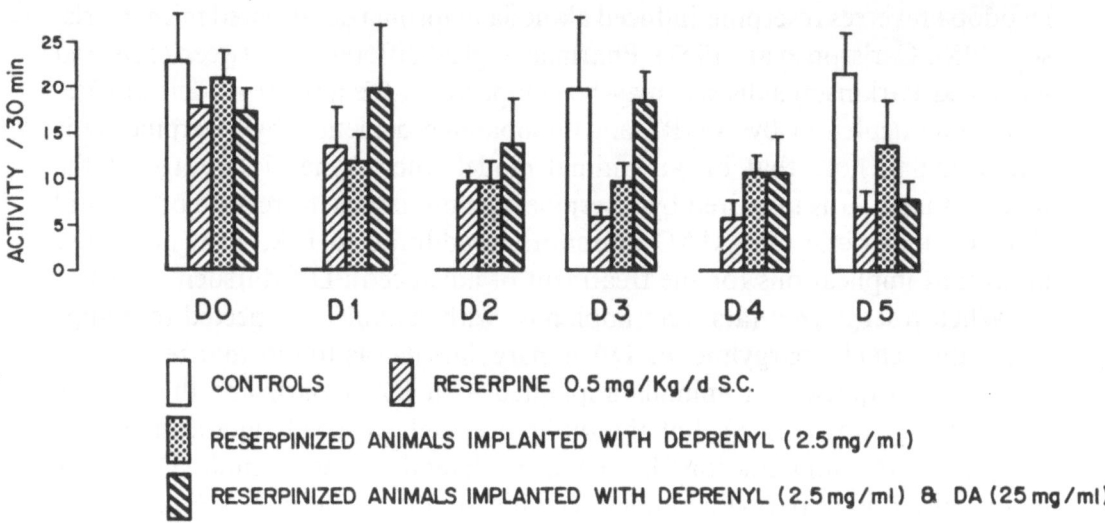

Fig. 11. Locomotor activity in controls and reserpinized rats after infusion of vehicle. Deprenyl (25 mg/ml) and DA (25 mg) + D. *D 0–D 5* Counts per 30 minute period obtained at baseline and up to day 5 of infusion

Obeso (personal communication) after chronic subcutaneous infusion of lisuride in humans.

Scatchard analysis of specific binding of ^{3}H-spiperone to membranes obtained from the striata ipsilateral to the infused ventricle, revealed that chronic reserpine treatment induced a 33% reduction in B_{max} values and a 55% reduction in Kd (Fig. 10). Infusion of DA agonists in reserpinized rats caused small changes in B_{max} and a further reduction of Kd values in DA-(33%), lisuride-(40%) and pergolide-(11%) infused rats.

The anti-akinetic effect of ICV infusions of DA + deprenyl was similar to DA + pargyline, but deprenyl was more than 10 times as potent as pargyline (Fig. 11). At the end of the infusion period, DA + deprenyl lost its anti-akinetic effect. This may be due to metabolism of deprenyl and reduction of brain level of deprenyl, preventing intraneuronal DA storage. This possibility is supported by the presence of high DOPAC levels in DA + deprenyl treated animals indicating that MAO inhibition was inefficient at the end of the experimental period (Fig. 12). As with DA + pargyline treated animals, DA + deprenyl infusion reversed reserpine-induced depletion of DA which surpassed normal levels. However, in animals infused with DA + deprenyl, reserpine-induced depletion of NE in the nucleus accumbens and hypothalamus was not reversed (Fig. 12).

DA infusions in MPTP-treated monkeys

Intact animals implanted with a pump delivering 0.1 ml of artificial CSF into the left ventricle were observed for a period of three weeks and found to have normal behavior and no side effects. After MPTP treatment, monkeys became

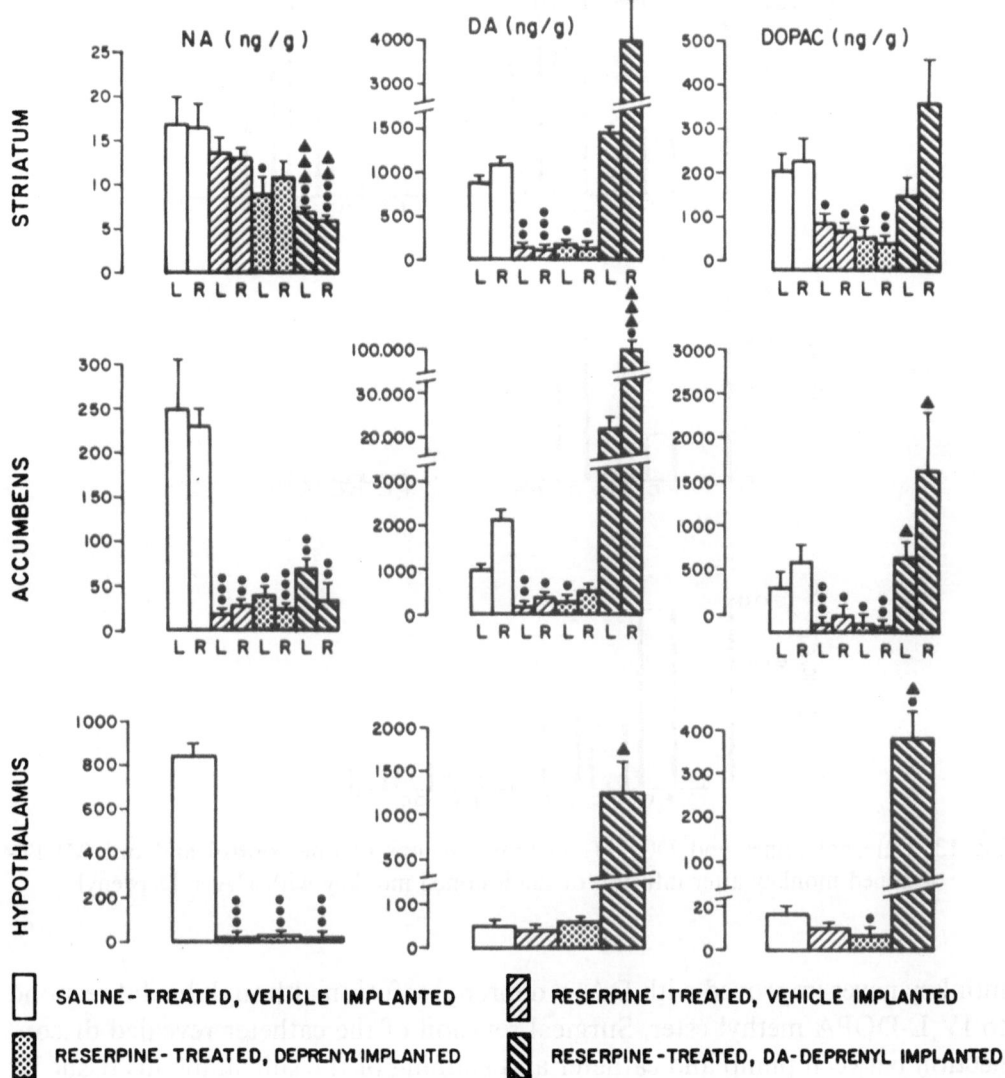

Fig. 12. Catecholamines and DOPAC in brain regions of control and reserpinized rats infused with vehicle, Deprenyl (2.5 mg/ml) and DA (25 mg) + D. Animals were implanted with a catheter into the right cerebral ventricle. *L* Left side, *R* right side. ● $p < 0.05$, ●● $p < 0.01$, ●●● $p < 0.001$ respect to the control group, ▲ $p < 0.05$, ▲▲ $p < 0.01$, ▲▲▲ $p < 0.001$ respect to the reserpinized vehicle infused group

severely akinetic and unable to move, feed themselves or drink. They had to be fed by hand and to be hydrated with occasional intravenous infusions of saline. Two such ill monkeys were implanted with a catheter in the left ventricle connected to an Infusaid pump and another ill monkey was used as a control. Two more animals died in the acute period of MPTP administration.

In one of the animals treated with DA (25 mg/h) plus deprenyl (0.2 mg/h) the pump was inserted subcutaneously into the lumbar region. In the other monkey the pump was placed over the skin in the skull, protected by an specially prepared helmet that allowed direct access to the port of the pump. This second

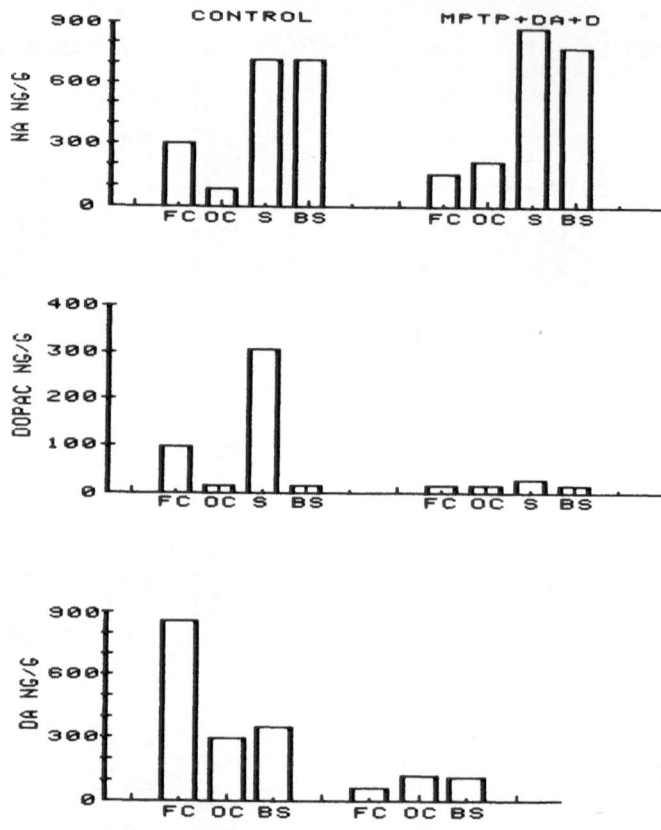

Fig. 13. Catecholamines and DOPAC in brain regions of one control and one MPTP-lesioned monkey after infusion of the lesioned monkey with DA + Deprenyl

monkey never improved with DA + deprenyl infusion although he did respond to IV L-DOPA methyl-ester. Surgical revision of the catheter revealed disconnection between pump and catheter and staining of the subcutaneous tissue by auto-oxidized DA metabolites. The first monkey had an increase in motor activity in less than 30 minutes after CSF infusion was substituted by DA and deprenyl. The animal regained almost normal behavioral activity, was able to drink and feed himself and moved normally, though slowly, around the cage. He had a tendency to rotate toward the contralateral side of the infused ventricle. This response was maintained for 9 days. Then, the skin under which the pump was placed became necrotic from ischemic pressure. The animal became slowly more akinetic. A surgical revision revealed disconnection of the pump. The animal was sacrificed under pentothal anesthesia.

Biochemical analysis of the brain revealed similar NE levels in both the MPTP treated and the control animal (Fig. 13). These findings are in agreement with other studies that showed that NE neurons are not affected by MPTP unless the animals are old, and receive repetitive treatment (Forno et al., 1986). DOPAC levels were greatly decreased in frontal cortex and striatum of the MPTP-treated monkey. This reduction could be produced by a combination

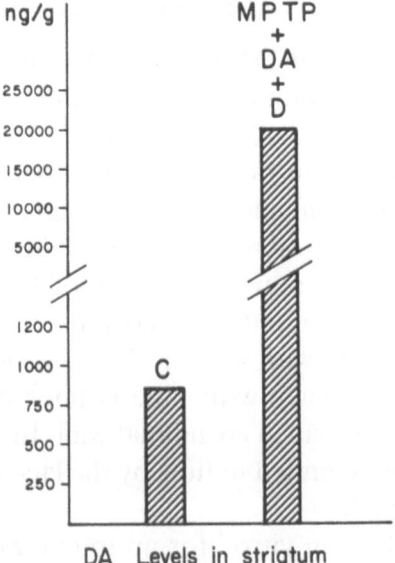

Fig. 14. DA levels in the striatum of one control monkey (*C*) and one MPTP lesioned monkey infused with DA and deprenyl into the ipsilateral ventricle

of effects: toxic effects of MPTP on DA neurons and inhibition of DA metabolism by deprenyl. DA levels were reduced in brain areas of the MPTP-treated monkey, but this depletion was reversed and DA levels in the striatum even surpassed normal levels with DA + deprenyl infusion (Fig. 14). This finding suggests that in large animals such as humans it may be possible to reverse neurotransmitter deficits by directly delivering the appropriate chemical to the selected target area and that the diffusion of the drug to other CNS structures may be limited. This approach may avoid adverse affects caused by the pharmacological action of DA and DA agonists on other DA pathways. Biochemical studies in brains of patients with PD have shown that the severity of the depletion of DA is greater in the nigrostriatal than in other DA pathways (Scatton et al., 1982, 1983). Some authors have also suggested that psychiatric side effects of levodopa therapy may be caused by the pharmacological action of levodopa in the limbic system (Goldstein et al., 1982). If this is the case, then delivery of DA agonists into the striatum through stereotaxic implanted catheters may be able to help these patients.

Morphological studies in implanted animals

Gross anatomy of brain and brain weight of implanted animals was normal. Microscopical examination and occasionally macroscopical observation of DA-infused animals revealed the presence of a black pigment that could be observed with hemotoxylin and eosin stain and was made more apparant by silver staining. The pigment was localized in the ventricular wall, the track of the cannulae, and in the meninges.

Conclusions

ICV infusion of DA, agonists or DA plus MAO inhibitors, reverses behavioral and biochemical abnormalities found in certain animals models of PD, namely 6-OH-DA unilateral lesion and reserpine-induced DA depletion in rats and MPTP-induced parkinsonism in monkeys. The observation period in rodents is limited by the time of infusion allowed by the implanted pump. In this period of time the intensity of behavioral responses elicited by DA infusion decreased. This fact can be explained by down regulation of DA receptors and, in fact, we observed a reduction in the number of ^{3}H-spiperone binding sites in animals with 6-OH-DA lesioned rats with chronic DA infusion. Another alternative explanation for decrease efficacy with time is poor diffusion of DA or DA agonists through a pigment-loaded ependimal wall. In MPTP-treated monkeys the success of the therapy is limited in time by the lack of an appropriate pump to fit the size of the animal.

DA, despite its stability "in vitro" for up to one week at 37 °C, is unstable "in situ". Auto-oxidation and pigment formation are probably increased by the buffering effect of CSF. Pharmacological effects of DA agonists at the maximal concentrations allowed by their solubility in water are weaker than those of DA. The use of more soluble DA agonists or precursors of DA is likely to solve this problem. Intracerebral infusion of specific neurotransmitters into selective target areas may reduce the frequency of side effects in patients with degenerative disorders of the CNS.

Acknowledgement

This work was supported by a grant from the Parkinson's Disease Foundation and by CAICYT grant 755/81.

The authors thanksfully acknowledge assistance and suggestions provided by Lee Voulters, M.D., Susan Lovelle, M.S. (Columbia University, New York), Jose Carlos Bustos, M.D., Carlos Magariños, Ph.D., and Antonio Parreño, M.D. (Centro "Ramon y Cajal", Madrid) during the performance of surgical procedures and behavioral studies. Armando Martinez, M.D. (Hospital Clinico de S. Carlos, Madrid) kindly performed histological studies. Jose Maria Carnero (Centro "Ramon y Cajal") performed the art work and Monica Fey Hatten kindly typed the manuscript.

Lisuride was provided by Schering A.G., Spain, and pumps used in primate studies by Infusaid, Inc., Milford, Massachusetts, U.S.A.

References

Carlsson A (1958) The occurrence, distribution and physiological role of catecholamines in the nervous system. Pharmacol Rev 10: 490–493
Carlsson A, Lindqvist M (1973) Effect of ethanol on the hydroxylation of tyrosine and tryptophan in rat brain in vivo. 25: 437–440
Carlsson A, Lindqvist M, Magnusson T (1957) 3,4-dyhydroxyphenylalamine and 5-hydroxytryptophan as reserpine antagonists. Nature 180: 1200
Fahn S, Calne DB (1978) Considerations in the management of parkinsonism. Neurology 28: 5–7

Forno LS, Langston JW, Delanney LE, Irwin I, Ricaurte GA (1986) Locus coeruleus lesions and eosinophylic inclusions in MPTP-treated monkeys. Ann Neurol 20: 449–455

Goldstein M, Lieberman A, Pearson J (1982) Relatively high levels of dopamine in the nucleus accumbens of levodopa treated patients with Parkinson's disease. J Neural Transm 54: 129–134

Granerus AK, Carlsson A, Svanborg A (1979) The aging neuron. Influence on symptomatology and therapeutic response in Parkinson's syndrome. In: Poirier LJ, Sourkes TL, Bedard PJ (eds) The basal ganglia [Advances in Neurology, vol 24]. Raven Press, New York, pp 327–334

Juncos JL, Mouradian MM, Fabbrini G, Seratti C, Chase TN (1986) Levodopa methylester treatment of Parkinson's disease. Neurology 36 [Suppl 1]: 218

Lesser RP, Fahn S, Snider SR, Cote LJ, Isgreen WP, Barrett RE (1979) Analysis of the clinical problems in parkinsonism and the complications of long-term Levodopa therapy. Neurology 29: 1253–1260

Marion MH, Stocchi F, Quinn NP, Jenner P, Marsden CD (1986) Repeated levodopa infusions in fluctuating Parkinson's disease: clinical and pharmacokinetic data. Clin Neuropharmacol 9: 165–181

Marsden CD (1980) On-off phenomena in Parkinson's disease. In: Rinne VK, Klinger M, Stam G (eds) Parkinson's disease. Current progress, problems, and management. Elsevier North Holland Bio-medical Press, Amsterdam New York, pp 241–254

Marsden CD, Parkes JD (1977) Success and problems of long-term levodopa therapy in Parkinson's disease. Lancet 1: 345–349

Marsden CS, Parkes JD, Quinn N (1982) Fluctuations of disability in Parkinson's disease—clinical aspects. In: Marsden CD, Fahn S (eds) Movement disorders. London, Butterworth, pp 96–122

Mena MA, Muradas V, Reiriz J, Bazan E, de Yebenes JG (1986) Pharmacokinetics of L-DOPA in patients with Parkinson's disease. In: Yahr MD, Bergmann J (eds) Parkinson's disease [Advances in Neurology, vol 45]. Raven Press, New York, pp 481–486

Muradas V, Bazan E, Gervas JJ, Mena MA, de Yebenes JG (1986) The effect of single and repeated administration of bromocriptine on monoamine metabolism in rat brain and ^{3}H-spiroperidol binding to striatal membranes. Movement Disorders 1: 103–112

Nutt JG, Woodward WR, Hammerstad JP, Carter JH, Anderson JL (1984) The on-off phenomenon in Parkinson's disease. Relation to levodopa absorption and transport. N Engl J Med 310: 483–488

Nutt J, Woodward WR (1986) Levodopa pharmacokinetics and pharmacodynamics in fluctuating parkinsonian patients. Neurology 36: 739–744

Obeso JA, Luquin MR, Martinez-Lage JM (1986) Intravenous lisuride corrects oscillations of motor performances in Parkinson's disease. Ann Neurol 19: 31–35

Quinn N, Marsden CD, Parkes JD (1982) Complicated response fluctuations in Parkinson's disease: response to intravenous infusion. Lancet 2: 412–413

Quinn N, Parkes JD, Marsden CD (1984) Control of on-off phenomenon by continuous intravenous infusion of levodopa. Neurology 34: 1131–1136

Reavill CH, Jenner P, Marsden CD (1983) Differentiation of dopamine agonists using drug-induced rotation in rats with unilateral or bilateral 6-OH-DA destruction of ascending dopamine pathways. Biochem Pharmacol 32: 865–870

Reches A, Fahn S (1982) 3-O-Methyldopa blocks DOPA metabolism in rat corpus striatum. Ann Neurol 12: 267–271

Reches A, Mielke LR, Fahn S (1982) 3-O-Methyldopa inhibits rotations induced by levodopa in rats after unilateral destruction of the nigrostriatal pathway. Neurology 32: 887–888

Rinne VK (1981) Treatment of Parkinson's disease: problems with a progressing disease. J Neural Transm 51: 161–174

Scatton B, Javoy-Agid F, Rouquier L, Dubois B, Agid Y (1983) Reduction of cortical dopamine, noradrenaline, serotonin and their metabolites in Parkinson's disease. Brain Res 275: 321–328

Scatton B, Rouguier L, Javoy-Agid F, Agid Y (1982) Dopamine deficiency in the cerebral cortex in Parkinson's disease. Neurology 32: 1039–1040

Seeman P, Chan-Wong M, Tedesco J, Wong K (1975) Brain receptors for antipsychotic drugs and dopamine direct binding assays. Proc Natl Acad Sci USA 72: 4376–4380

Vanderwende C (1964) Studies on the oxidation of dopamine to melanine by rat brain. Arch Int Pharmacodyn 152: 433–444

Authors' address: J. G. de Yebenes, M.D., Servicio de Neurologia, Hospital Universitario "S Carlos" Planta 2° Sur, C Martin Lagos, Ciudad Universitaria, E-28040 Madrid, Spain.

J Neural Transm (1988) [Suppl] 27: 161–175
© by Springer-Verlag 1988

Receptor changes during chronic dopaminergic stimulation

P. Jenner, S. Boyce, and **C. D. Marsden***

MRC Movement Disorders Research Group, University Department of Neurology and Parkinson's Disease Society Research Centre, Institute of Psychiatry and King's College Hospital Medical School, London, U.K.

Summary. The underlying cause of the long term complications of L-DOPA or dopamine agonist therapy in Parkinson's disease remains unknown. Previous studies of repeated administration of L-DOPA or bromocriptine to rodents have shown increases, decreases or no change in brain dopaminergic activity. For this reason we have re-examined the effects of chronic L-DOPA or dopamine agonist administration on brain dopamine receptor function in rats.

Repeated intraperitoneal administration of L-DOPA to rats for 21 days followed by 3 days drug withdrawal caused an enhancement of apomorphine-induced stereotypy but no apparent alteration in striatal dopamine receptor numbers or affinity (as judged by ^{3}H-spiperone; ^{3}H-NPA and ^{3}H-piflutixol binding).

Chronic oral administration of L-DOPA plus carbidopa to rats for one year was without effect on apomorphine-induced stereotypy or striatal D-2 dopamine receptors. Similarly, no effects were observed on striatal dopamine function following one year's administration of bromocriptine. Pergolide produced an enhancement of apomorphine-induced stereotypy but a decrease in D-2 receptor density as judged by ^{3}H-spiperone binding.

In rats with a unilateral 6-OHDA lesion of the medial forebrain bundle the oral administration of L-DOPA plus carbidopa for 4 weeks, followed by 4 days withdrawal, enhanced the rate of apomorphine-induced contraversive rotation.

It appears difficult, at least in rats, to manipulate striatal dopamine receptors with L-DOPA or dopamine agonist drugs. An enhancement of motor behaviour can occur in the presence of no change or a decrease in dopamine receptor numbers identified by in vitro ligand binding to tissue homogenates.

Introduction

The treatment of Parkinson's disease with L-DOPA is complicated by loss of the initial benefit after some years of continued therapy. The "wearing-off" of

---

* Present address: University Department of Clinical Neurology, The National Hospital for Nervous Diseases, Queen Square, London, WC1N 3BG, U.K.

L-DOPA is accompanied by the onset of a number of other long-term complications of therapy including "on-off" phenomena, peak dose and biphasic dyskinesias, early morning dystonia and "yo-yoing" (Marsden, 1980; Marsden and Parkes, 1976, 1977; Marsden et al., 1982). The reasons for the loss of effect of L-DOPA and onset of other motor complications is not clear. It has been suggested that they may represent a pharmacokinetic change in the handling of L-DOPA, or a pharmacodynamic effect due to altered dopamine receptor sensitivity caused by prolonged agonist bombardment, or a progression of the underlying pathology of Parkinson's disease itself. Whether or not the chronic use of dopamine agonist drugs, such as bromocriptine or pergolide, in the treatment of Parkinson's disease leads to similar problems is not clear. It has been claimed that chronic bromocriptine therapy may not be associated with some of these long-term difficulties (Stern and Lees, 1983; Rinne, 1987) but this requires replication.

In classical pharmacological terms, continuous application of an agonist to its receptor would be expected to lead to desensitisation or down-regulation (Trendelenburg, 1966). In contrast, deprivation of the receptor of its transmitter by denervation or by blockade should cause supersensitivity or up-regulation. Brain dopamine receptors appear to react in this manner in response to treatment with antagonists. Thus, the repeated administration of neuroleptic drugs has been consistently shown to cause an increased behavioural supersensitivity to applied dopamine agonists, which correlates with an enhanced number of receptors as measured using ligand binding assays (see Muller and Seeman, 1978). However, no consistent picture has yet emerged of the effects of repeated administration of dopamine agonist compounds on striatal dopamine function.

Previous studies of repeated administration of L-DOPA, bromocriptine, or pergolide

A number of different studies of repeated administration of L-DOPA have been undertaken in recent years using a variety of rodent species (Table 1). Those studies examining dopamine mediated behaviours have reported either an enhancement of behavioural response to agonist administration or no change; so far there have been no reports of a decreased behavioural response to apomorphine.

The results of investigations examining changes in dopamine receptors, as judged by ligand binding studies or estimates of adenylate cyclase activity, are even more conflicting. There are reports of an increase, a decrease, or no change in striatal dopamine receptor function as a result of repeated L-DOPA administration; the majority opt for no change.

Repeated bromocriptine administration has been studied for only brief periods (Table 2). Bromocriptine treatment has always resulted in a change in the behavioural response to dopamine agonists, but both an increase and a decrease in the response have been reported. However, most studies have reported no change in dopamine receptors, although two investigations showed an increase in parameters associated with dopamine receptor function.

Table 1. Summary of previous studies on the effects of repeated administration of L-DOPA on brain dopamine receptor function

Species	Number of studies			
	Guinea-pig	Mouse		Rat
	3	6		11
Route	Oral	IP	IM	SC
	16	2	1	1

	Protocol employed		
	Dose range (mg/kg/day)	Duration	Withdrawal
L-DOPA	20–9,000	10 d–22 m	0–14 d
L-DOPA plus carbidopa	40–500 10–100	4–21 d	0–4 d

	Results obtained		
	Increase	Decrease	No change
Behavioural sensitivity	7/11	0/11	4/11
Biochemical sensitivity	3/13	2/13	8/13

The data summarizes the protocols and results obtained in previous studies of chronic L-DOPA administration

Data accumulated from: Bailey et al., 1979; Friedhoff et al., 1977; Globus et al., 1982; Gudelsky et al., 1975; Hall et al., 1984; Jackson et al., 1983; Klawans et al., 1977; Klawans et al., 1979; List and Seeman, 1979; Mishra et al., 1978; Papavasiliou et al., 1981; Ponzio et al., 1984; Pycock et al., 1982; Reches et al., 1982; Sahakian et al., 1980; Smith et al., 1979; Weiner et al., 1981; Wilner et al., 1980

So far only two studies on the effects of repeated pergolide administration have been carried out. In rats, repeated pergolide administration decreased the number of striatal dopamine receptors identified by ^{3}H-NPA (Fuxe et al., 1981), while in guinea pigs an enhancement of apomorphine-induced stereotypy was observed (Klawans et al., 1981).

Clearly there has been a marked disparity in the results arising from studies of repeated administration of dopamine agonist drugs. Many reasons may be put forward to explain such discrepancies. These would include differences in the species employed, in the route of administration of drugs, and in the withdrawal period (if any), as well as in the manner in which the behavioural and

Table 2. Summary of previous studies on the effects of repeated administration of bromocriptine on brain dopamine receptor function

	Number of studies		
Species	Guinea-pig	Mouse	Rat
	1	1	7

Route	Oral	IP	IM	SC
	3	3	0	3

	Protocol employed		
	Dose range (mg/kg/day)	Duration	Withdrawal
Bromocriptine	2–35	5–30 d	1–7 d

	Results obtained		
	Increase	Decrease	No change
Behavioural sensitivity	2/3	1/3	0/3
Biochemical sensitivity	0/6	2/6	4/6

Data accumulated from: Globus et al., 1982; Goldstein et al., 1978; List and Seeman, 1979; Mishra et al., 1978; Nausieda et al., 1978; Quikard Iversen et al., 1978; Smith et al., 1979

biochemical changes were measured. There has been so little consistency that it is difficult to compare studies from different laboratories.

The experiments to be described represent a further attempt to determine whether or not the administration of L-DOPA, bromocriptine or pergolide causes either behavioural or biochemical changes in striatal dopamine receptor function in rats.

Short-term intraperitoneal administration of L-DOPA followed by withdrawal of drug

Because of the lack of agreement on the effects of repeated administration of L-DOPA on dopamine function in the striatum, we returned to the original protocol utilised by Klawans and his colleagues (Klawans et al., 1977). These authors showed that behavioural supersensitivity to dopamine agonists occurred after administration of L-DOPA.

L-3,4-Dihydroxyphenylalanine (L-DOPA) was administered to rats for 14

Table 3. The effect of repeated administration of L-DOPA (40 or 200 mg/kg/day ip) for 14 days and subsequent withdrawal (3 days) on specific ^{3}H-spiperone (0.1–4.0 nM), ^{3}H-N, n-propylnorapomorphine (0.05–2.0 nM) or ^{3}H-piflutixol (0.08–1.8 nM)

Ligand	Saline		L-DOPA (40 mg/kg)		L-DOPA (200 mg/kg)	
	B_{max}	K_D	B_{max}	K_D	B_{max}	K_D
^{3}H-spiperone	25.9 ± 1.0	0.14 ± 0.01	27.6 ± 0.8	0.14 ± 0.01	26.5 ± 1.0	0.14 ± 0.02
^{3}H-N, n-propyl-norapomorphine	16.4 ± 1.3	0.96 ± 0.04	16.8 ± 0.3	0.97 ± 0.02	18.7 ± 0.9	0.97 ± 0.03
^{3}H-piflutixol	120 ± 5	0.26 ± 0.05	138 ± 9	0.31 ± 0.06	133 ± 8	0.28 ± 0.05
+ sulpiride	94 ± 3	0.33 ± 0.05	105 ± 4	0.35 ± 0.06	103 ± 5	0.36 ± 0.04

The results are expressed as mean (± 1 S.E.M.) of the values obtained from Scatchard analysis of data from 4–5 separate tissue pool for each treatment group, each of 6 ligand concentrations being examined in duplicate or triplicate. Specific binding of ^{3}H-spiperone and ^{3}H-N, n-propylnorapomorphine was defined using (+)-butaclamol (10^{-6} M) and (+)-ADTN (10^{-6} M) respectively. Specific binding of ^{3}H-piflutixol to D-1 and D-2 sites was defined using *cis*-flupenthixol (10^{-6} μM) alone and to D-1 sites alone and in the presence of (−)-sulpiride 3×10^{-5} M). B_{max} is expressed in pmoles/g wet weight of tissue and K_D in nM. The variation in non-specific binding in control treated animals was less than 8%

days. Three days after withdrawal of the drug, alterations in behaviour were examined, together with changes in the populations of D-1 and D-2 receptors in the striatum (Hall et al., 1984).

Assessment of stereotyped behaviour, induced by apomorphine (0.03–2.0 mg/kg, s.c., 15 min previously), showed the dose-response curves for stereotypy to be shifted to the left by treatment with L-DOPA in a dose-dependent manner. In addition, more than 50% of animals receiving the largest dose of L-DOPA exhibited occasional explosive motor behaviour that was not apparent in the control group. However, no changes in the number of specific binding sites for ^{3}H-spiperone, ^{3}H-N,n-propylnorapomorphine (NPA), or ^{3}H-piflutixol was found in tissue preparations of striatum (Table 3). Similarly, basal and dopamine-stimulated activity of adenylate cyclase in the group treated with L-DOPA did not differ from that found in the control animals.

The conclusion from these studies was that L-DOPA treatment for 14 days, followed by 3 days of drug withdrawal, induced apparent behavioural sensitization of dopamine receptors in the brain, but not the receptor itself at least as measured in vitro.

Administration of L-DOPA, bromocriptine or pergolide continuously for 12 months

In the previous study, down-regulation of dopamine receptors might not have occurred following L-DOPA because only a relatively short period of drug administration was employed. For this reason, a study was carried out in which rats were treated with L-DOPA (153–173 mg/kg/day) plus carbidopa (19–22 mg/kg/day) for a period of 1 year. However, despite this dose being sufficient to reduce the growth rate of the rats by some 20–25%, (compared to control animals), the ability of apomorphine to induce stereotyped behaviour over the period of 1 year did not change, and there was no alteration in the binding of ^{3}H-spiperone to D-2 receptors (Table 4).

In normal rats it appears to be difficult to produce a down-regulation of D-2 sites using chronic administration of L-DOPA. In parallel studies, the dopamine agonist drugs bromocriptine (3.3–4.6 mg/kg/day) or pergolide (0.36–0.51 mg/kg/day) were administered for 1 year. Again, no effects were observed after administration of bromocriptine. However, pergolide did produce an enhancement of apomorphine-induced stereotyped behaviour, but there was a decrease in ligand binding to D-2 receptors. This combination of effects is difficult to reconcile.

Involvement of other neurotransmitter systems

So far, consideration has only been given to alterations in brain dopamine function as a consequence of chronic L-DOPA or dopamine agonist therapy. However, L-DOPA, bromocriptine and pergolide also act to alter brain noradrenaline and 5 HT function. Also, other neuronal systems are likely to be altered indirectly as a result of drug action on monoamine receptors. For

Table 4. The effect of continuous administration to rats for 12 months of L-DOPA (153–173 mg/kg/day) plus carbidopa (19–22 mg/kg/day) or bromocriptine (3.3–4.6 mg/kg/day) or pergolide (0.36–0.51 mg/kg/day) on apomorphine-induced stereotyp and ^{3}H-spiperone binding to striatal membrane

Drug treatment	ED_{50} mg/kg. s.c. apomorphine-induced stereotypy	B_{max} (pmol/g) tissue ^{3}H-spiperone binding
Control	0.15 ± 0.02	20.0 ± 0.5
L-DOPA plus carbidopa	0.15 ± 0.03	17.6 ± 2.0
Bromocriptine	0.13 ± 0.02	17.9 ± 2.0
Pergolide	$0.06 \pm 0.02*$	$13.6 \pm 0.7*$

Dose-response curves of apomorphine- (0.0625–1.0 mg/kg s.c.) induced stereotyped behaviour for each experimental group were tested for differences using ANOVA; when significant F ratios were obtained. ED_{50} values were determined using curve-fitting analysis. Values are expressed as mean ($\pm$ 1 S.E.M.). 6 animals were used at each dose for each experimental group

Binding is expressed as mean ($\pm$ 1 S.E.M.) of the values obtained from Eadie-Hofstee analysis from three separate tissue pools for each experimental group, each of 6 concentrations of ligand being examined in triplicate. Specific bind of ^{3}H-spiperone was defined using ($\pm$) sulpiride (10^{-5} M)

* $p < 0.05$ compared to control using Student's t-test

example, one year's treatment with L-DOPA plus carbidopa, alters some neuropeptides (de Ceballos et al., 1986).

Administration of carbidopa for 12 months did not alter [Met[5]]- or [Leu[5]]enkephalin levels in the striatum, or neurotensin levels in the striatum, nucleus accumbens or substantia nigra (Table 5). In contrast, administration of L-DOPA plus carbidopa for 12 months decreased the [Met[5]]-, but not [Leu[5]]enkephalin, content in striatum. Also, L-DOPA plus carbidopa treatment decreased the neurotensin level in striatum and substantia nigra, but not the nucleus accumbens. Similarly, 12 months administration of pergolide reduced [Met[5]]-, but not [Leu[5]]enkephalin, levels in striatum, and only decreased neurotensin content in striatum. In contrast, administration of bromocriptine for a 12 month period had no effect on [Met[5]]- or [Leu[5]]enkephalin levels in striatum. While bromocriptine reduced neurotensin levels in the striatum there was no effect in the nucleus accumbens, and the neurotensin content of substantia nigra increased.

Administration of L-DOPA to rats lesioned with 6-OHDA

The experiments so far described were carried out in normal rats, whereas the interest in the clinical effects of L-DOPA is in Parkinson's disease in which there is pronounced neuronal loss of dopamine in brain. Three previous studies have been carried out using the rotating rodent model (Table 6). The single behavioural study suggests no change in striatal function and the two bio-

Table 5. Effect of continuous administration of L-DOPA, bromocriptine and pergolide for 12 months on [Met⁵]- and [Leu⁵]enkephalin and neurotensin content in basal ganglia

Treatment	[Met⁵]enkephalin (pmol/g tissue) Striatum	[Leu⁵]enkephalin (pmol/g tissue) Striatum	Neurotensin (pmol/g tissue)		
			Striatum	Nucleus accumbens	Substantia nigra
Control	62 ± 6	14 ± 2	1.1 ± 0.1	3.4 ± 0.8	2.6 ± 0.2
Carbidopa	56 ± 6	15 ± 2	0.8 ± 0.0	2.9 ± 0.3	3.4 ± 0.4
L-DOPA plus carbidopa	37 ± 5*	13 ± 3	ND	3.7 ± 0.2	0.4 ± 0.2*
Bromocriptine	63 ± 5	15 ± 2	ND	2.8 ± 0.3	4.0 ± 0.2*
Pergolide	30 ± 4*	12 ± 3	ND	2.4 ± 0.2	2.3 ± 0.2

Animals were treated continuously for 12 months with drugs dissolved in drinking water. Actual intake of drug was as follows: carbidopa alone (26.0–28.1 mg/kg per day), L-DOPA (123–124 mg/kg per day) plus carbidopa (15.4–15.5 mg/kg per day), bromocriptine (3.6–4.2 mg/kg per day) and pergolide (0.40–0.48 mg/kg per day). The peptides were measured with selective radioimmunoassays after HPLC separation of the brain extracts. Results are mean ± 1 S.E.M. of 5–6 animals

* $p < 0.05$ versus age-matched controls

ND Not detectable

Table 6. Effect of L-DOPA administration on behaviour and biochemical parameters in rodents with a prior unilateral 6-hydroxydopamine lesion of the medial forebrain bundle

Species	Dose of L-DOPA	Route	Duration	With-drawal	Circling behaviour	Ligand binding	B$_{max}$
Mouse	500–600 mg/kg/day	p.o.	2 months	0–7 day	L-DOPA, no change induced	—	—
Rat	200 mg/kg/day	p.o.	30 days	24 hours	—	[³H]-spiperone	21% lesion; no change nonlesion
Rat	200 mg/kg/day + 20 mg/kg/day carbidopa	p.o.	14 day	3 day	—	[³H]-spiperone	Decrease (not B$_{max}$) (both lesion and nonlesion)

Data taken from: Pycock and Marsden, 1977; Suga, 1980; Reches et al., 1984

chemical investigations suggest a down regulation of dopamine receptors. Consequently, the effects of administration of L-DOPA to rats with a unilateral lesion induced by 6-hydroxydopamine (6-OHDA) were re-examined.

Male Wistar rats received a unilateral lesion of the medial forebrain bundle induced with 6-hydroxydopamine (8 µg). Four-6 weeks later they were tested for their response to the administration of apomorphine (0.5 mg/kg, s.c., 20 min previously). Those animals exhibiting contraversive rotation at the rate of more than five turns per minute were included in the groups treated with drug. Such animals received either carbidopa (13.5–20.9 mg/kg/day) or L-DOPA (82.8–167 mg/kg/day) plus carbidopa (10.0–21.0 mg/kg/day) for a period of 1 month. At the end of this period, the animals were allowed a withdrawal period of 3 or 4 days, prior to behavioural testing.

During the course of the administration of drug, animals receiving treatment with L-DOPA showed spontaneous contraversive rotation for the first week of administration of drug, but this was not observed subsequently. When assessed after withdrawal of drug, the animals showed no spontaneous circling behaviour. However, administration of a range of doses of apomorphine produced an enhanced contraversive circling response in the animals treated with L-DOPA, compared with those animals receiving carbidopa alone. There was no difference in the stereotyped response of the animals to administration of apomorphine, nor were there any changes in the degree of postural asymmetry. The rate of contraversive rotation to apomorphine (0.5 mg/kg/day, s.c., 20 min previously) in the animals receiving L-DOPA was markedly enhanced compared with the rate of rotation prior to the start of administration of drug (before, 20.0 ± 1.2 per 2 min period; n = 42; after, 40.7 ± 7.1 per 2 min period; n = 6; $p < 0.05$). However, in the group treated with carbidopa there was no change in the rate of rotation over the period of the experiment (before, 21.4 ± 1.6 per 2 min period; n = 41; after, 24.8 ± 2.3 per 2 min period; n = 6; $p > 0.05$). At the end of the behavioural experiment, the animals were killed for biochemical analysis; there was a greater than 90% loss of the content of dopamine in the striatum on the lesioned side of the animals treated with 6-hydroxydopamine.

The results of the treatment of animals with a prior lesion with 6-hydroxydopamine suggest that the administration of L-DOPA can cause an exaggeration of the behavioural response to a dopamine agonist. This contrasts with the previous report of Pycock and Marsden (1977) in which no change in the circling rate of mice was observed. At present these findings cannot be reconciled with the previous biochemical studies, showing that repeated administration of L-DOPA to rats with unilateral nigrostriatal lesions caused a decrease in the number of specific binding sites for ^{3}H-spiperone. However, in these studies, no behavioural observations on the rate of rotational response were reported. It may be that alteration in the rate of rotation can result from a change in the balance of dopamine function in the lesioned and intact hemispheres. Thus, Reches and colleagues found L-DOPA to decrease the number of binding sites for ^{3}H-spiperone in both striata (Reches et al., 1984).

Dopamine receptors in Parkinson's disease and in primates treated with MPTP

The overall picture emerging from the rodent experiments still does not provide a clear indication of the effect of L-DOPA or dopamine agonist drugs on striatal dopamine receptors. Perhaps the models used are not valid or perhaps our current concept, of the nature of dopamine receptors in Parkinson's disease is incorrect. So what is the status of dopamine receptors in Parkinson's disease and how are they affected by drug treatment?

Individual studies in human patients suggest a picture equally confused to that in the rodent investigations. Reports of increases, decreases and no change in D-2 receptor density in Parkinson's disease have appeared (Lee et al., 1978; Reisine et al., 1977; Rinne et al., 1981). Some authors hold that striatal D-2 receptors are increased in Parkinson's disease and that this is particularly marked in untreated patients (see Seeman, 1987; Beaulieu, 1987). They contend that this increase may be reversed to normal by drug treatment, although this is contested (Bokobza et al., 1984). The density of D-2 receptors is not altered by duration of disease and an increased density of D-2 sites is observed in those who respond well to drug treatment, those who develop dyskinesias and "on-off" responses and in those severely disabled by the disease.

There does not appear to be a consistent dopamine receptor loss or "down-regulation" to explain loss of drug efficacy. Indeed, both the rodent studies and post-mortem investigations suggest that it is difficult to reduce D-2 receptor density by dopamine agonist treatment.

It has also become apparent that the density of D-2 receptors is not uniformly altered in Parkinson's disease. Thus, alterations in dopamine receptor density follow the severity of dopamine loss. Changes in D-2 receptors are more apparent in the putamen with less obvious alterations in the caudate nucleus or nucleus accumbens. This implies that the effects of drug treatment may be more apparent in putamen than elsewhere.

The study of dopamine receptors in post-mortem parkinsonian brain is complicated by drug treatment, duration of illness, severity of disease and many other factors. The ability of MPTP to induce selective nigral cell death and parkinsonism in primates provides an opportunity to determine the effect of denervation alone on the status of striatal dopamine receptors.

In common marmosets treated with MPTP 10 days or 4–6 weeks previously, dopamine levels in caudate-putamen were rendered by more than 90%, accompanied by a marked decrease in ^{3}H-dopamine uptake and histological evidence for nigral cell loss. However, despite the considerable degree of denervation induced by MPTP, there was no change in the number or affinity of D-2 receptors measured by specific ^{3}H-spiperone binding to caudate homogenates.

However, using autoradiographic analysis of ^{3}H-spiperone binding to slices of marmoset brain, selective increase in D-2 sites in ventrolateral striatum was demonstrated (Waters et al., 1985; see also, Joyce et al., 1985). This suggests that specific regional changes in D-2 receptor density do occur following MPTP treatment, but that these may not be detected using homogenate techniques.

Interestingly, L-DOPA therapy in MPTP-treated monkeys was shown to enhance the number of D-2 sites identified by specific ^{3}H-spiperone to striatal homogenates, whereas bromocriptine was less effective in this respect (Bedard et al., 1986). At this time it is difficult to reconcile these findings with what appears to occur in Parkinson's disease, as judged by post-mortem studies.

Conclusions

The evidence available from animal experiments suggests that dopamine agonists do not readily down regulate the brain dopamine receptor population. Both the post-mortem studies in Parkinson's disease and the experimental investigations in animals indicate that neither a loss of dopamine receptors nor a drug-induced decrease in receptor numbers can explain decreasing drug efficacy during the long term treatment of Parkinson's disease. Indeed, the success of intravenous L-DOPA infusions in controlling fluctuations in drug response in parkinsonian patients (Quinn et al., 1983) suggest that brain dopamine receptors remain responsive to dopamine if supplied continuously in the correct concentration.

The ability of L-DOPA and dopamine agonist drugs to enhance the behavioural response to an acute administration of apomorphine may have relevance to the onset of hyperkinetic disorders. However, there is no compelling experimental or post-mortem data to link this effect to any dopamine receptor changes. Indeed, the approach taken so far may be too narrow to detect any involvement of dopamine receptor alterations in L-DOPA-induced dyskinesias. Further studies should consider the following:

(1) Alterations in D-2 dopamine receptors may not be uniform throughout the caudate-putamen. Subregional studies using autoradiographic techniques may reveal selective changes in dopamine receptor populations.

(2) Alterations in dopamine receptors in brain areas other than caudate-putamen may occur as a result of Parkinson's disease or its treatment. For example, dopamine receptor populations in substantia nigra which control the firing of outflow pathways from basal ganglia may be involved.

(3) Alterations may occur in D-1 as well as D-2 receptors, and these may contribute to Parkinson's disease and its treatment.

(4) In rats, at least, a functional linkage exists between striatal D-1 and D-2 receptors that breaks down following denervation such that D-1 and D-2 systems function independently of one another. Whether such linkage occurs in human brain is not known nor the effect of Parkinson's disease on the interaction between these receptor sub-types.

(5) Systems other than dopamine receptor may be involved. The drugs used to treat Parkinson's disease act on several different neuronal systems. Also, the response of other neurones activated by dopamine receptor occupation may contribute to the dyskinesias and their response may be altered by the disease

process. In particular, GABA mediated outflow from caudate-putamen may play a critical role.

Acknowledgements

This work was supported by the Medical Research Council, The Parkinson's Disease Society and the Research Funds of the Bethlem Royal and Maudsley Hospitals and King's College Hospital.

References

Agid Y, Javoy-Agid F, Ruberg M (1987) Biochemistry of neurotransmitters in Parkinson's disease. In: Marsden CD, Fahn S (eds) Movement disorders, vol 2. Butterworths, London, pp 166–230

Bailey RC, Jackson DM, Bracs PV (1979) Long term L-DOPA pretreatment of mice: central subsensitivity or supersensitivity? Psychopharmacology 66: 55–61

Beaulieu M (1987) Clinical importance of D-1 and D-2 receptors. Can J Neurol Sci 14 [Suppl]: 402–406

Bedard PJ, Di Paolo T, Falardeau P, Boucher R (1986) Chronic treatment with L-DOPA, but not bromocriptine induces dyskinesia in MPTP-parkinsonian monkeys. Correlation with [^{3}H]spiperone binding. Brain Res 379: 294–299

Bokobza B, Ruberg M, Scatton B, Javoy-Agid F, Agid Y (1984) [^{3}H]spiperone binding, dopamine and HVA concentrations in Parkinson's disease and supranuclear palsy. Eur J Pharmacol 99: 167–175

De Ceballos M, Boyce S, Jenner P, Marsden CD (1986) Alterations in [Met^5]- and [Leu^5]enkephalin and neurotensin content in basal ganglia induced by long-term administration of dopamine agonist and antagonist drugs to rats. Eur J Pharmacol 130: 305–309

Friedhoff AJ, Bonnet K, Rosengarten H (1977) Reversal of two manifestations of dopamine receptor supersensitivity by administration of L-DOPA. Res Commun Chem Pathol Pharmacol 16: 411–423

Fuxe K, Agnati LF, Kohler C, Kuonen D, Ogren S-O, Anderson K, Hokfelt T (1981) Characterization of normal and supersensitive dopamine receptors: effects of ergot drugs and neuropeptides. J Neural Transm 51: 3–37

Globus M, Bonnet J, Lerer B, Belmaker RH (1982) The effect of chronic bromocriptine and L-DOPA on spiperone binding and apomorphine-induced stereotypy. Psychopharmacology 78: 81–84

Goldstein M, Lew JY, Nakamura S, Battista AF, Lieberman A, Fuxe K (1978) Dopaminophic properties of ergot alkaloids. Fed Proc 37: 2202–2206

Gudelsky GA, Thornburg JE, Moore KE (1975) Blockade of alpha-methyltyrosine-induced supersensitivity to apomorphine by chronic administration of L-DOPA. Life Sci 16: 1331–1338

Guttmann M, Seeman P, Reynolds GP, Riederer P, Jellinger K, Tourtellotte WW (1986) Dopamine D-2 receptors density remains constant in treated Parkinson's disease. Ann Neurol 19: 487–492

Hall MD, Cooper DR, Fleminger S, Rupniak NMJ, Jenner P, Marsden CD (1984) Behavioural and biochemical alterations in the function of dopamine receptors following repeated administration of L-DOPA to rats. Neuropharmacology 23: 545–553

Hantraye P, Loc'h C, Tacke U, Riche D, Stulzaft O, Doudet D, Guibert B, Naquet R, Maziere B, Maziere M (1986) "In vivo" visualisation by positron emission tomography of the progressive striatal dopamine receptor damage occurring in MPTP-intoxicated non-human primate. Life Sci 39: 1375–1382

Jackson DM, Jenkins OF, Malor R, Christie MJ, Gregory P (1983) Chronic L-DOPA treatment of rats and mice does not change the sensitivity of post-synaptic dopamine receptors. Naunyn Schmiedebergs Arch Pharmacol 324: 271–274

Joyce JN, Marshall JF, Bankiewicz KS, Kopin IJ, Jacobowitz DM (1985) Hemiparkinsonism in a monkey after unilateral internal carotid artery infusion of 1-methyl-4-phenyl-1,2,3,6,tetrahydropyridine (MPTP) is associated with regional ipsilateral changes in striatal dopamine D-2 receptor density. Brain Res 382: 360–364

Klawans HL, Goetz C, Nausieda PA, Weiner WJ (1977) Levodopa induced dopamine receptor hypersensitivity. Ann Neurol 2: 125–129

Klawans HL, Hitri A, Carvey PM, Nausieda PA, Weiner WJ (1979) Effect of chronic dopaminergic agonism on striatal membrane dopamine binding. Adv Neurol 24: 245–253

Klawans HL, Koller WC, Diamond BI, Nausieda PA, Weiner WJ (1981) Animal models in the evaluation of anti-parkinson agents. In: Clifford Rose F, Capildeo R (eds) Research progress in Parkinson's disease. Pitman Medical, London, pp 282–297

Lee T, Seeman P, Rajput A, Farley IJ, Hornykiewicz O (1978) Receptor basis for dopaminergic supersensitivity in Parkinson's disease. Nature 273: 59–61

List SJ, Seeman P (1979) Dopamine agonists reverse the elevated ^{3}H-neuroleptic binding in neuroleptic pre-treated rats. Life Sci 24: 1447–1452

Marsden CD (1980) "On-off" phenomena. In: Rinne UK, Klinger M, Stamm G (eds) Parkinson's disease—current progress, problems, and management. Elsevier North Holland, Amsterdam, pp 241–254

Marsden CD, Parkes JD (1976) "On-off" effects in patients with Parkinson's disease on chronic levodopa therapy. Lancet 1: 292–295

Marsden CD, Parkes JD (1977) Success and problems of long-term levodopa therapy in Parkinson's disease. Lancet 1: 345–349

Marsden CD, Parkes JD, Quinn N (1982) Fluctuations of disability in Parkinson's disease—clinical aspects. In: Marsden CD, Fahn S (eds) Movement disorders. Butterworth Scientific, London, pp 96–122

Mishra RK, Wong Y-W, Varmuza SL, Tuff L (1978) Chemical lesion and drug-induced supersensitivity and sub-sensitivity of caudate dopamine receptors. Life Sci 24: 443–446

Muller P, Seeman P (1978) Dopaminergic supersensitivity after neuroleptics. Time course and specificity. Psychopharmacology 60: 1–11

Nausieda PA, Weiner WJ, Kanapa DJ, Klawans HL (1978) Bromocriptine-induced behavioral hypersensitivity: implications for the therapy of parkinsonism. Neurology 28: 1183–1188

Papavasiliou PS, Miller ST, Thal LJ, Nerder LJ, Houlihan G, Rao SN, Stevens JM (1981) Age-related motor and catecholamine alterations in mice on levodopa supplemented diet. Life Sci 28: 2945–2952

Ponzio F, Cimino M, Achilli G, Lipartiti M, Perego C, Vantini G, Algeri S (1984) In vivo and in vitro evidence of dopaminergic system down regulation induced by chronic L-DOPA. Life Sci 34: 2107–2116

Pycock C, Dawbarn D, O'Shaughnesy C (1982) Behavioural and biochemical changes following chronic administration of L-DOPA to rats. Eur J Pharmacol 79: 201–215

Pycock C, Marsden CD (1977) Central dopaminergic receptor supersensitivity and its relevance to Parkinson's disease. J Neurol Sci 31: 113–121

Quik M, Iversen LL (1978) Subsensitivity of the rat striatal dopaminergic system after treatment with bromocriptine: effects of ^{3}H-spiperone binding and dopamine stimulated cyclic AMP formation. Naunyn Schmiedebergs Arch Pharmacol 304: 141–145

Quinn NP, Parkes D, Marsden CD (1983) The control of clinical fluctuations in Parkinson's disease by continuous intravenous administration of levodopa. Neurology 33 [Suppl 2]: 90–91

Reches A, Wagner HR, Jackson-Lewis V, Yablonskaya-Alter E, Fahn S (1984) Chronic levodopa or pergolide administration induces down regulation of dopamine receptors in denervated striatum. Neurology 34: 1208–1212

Reches A, Wagner HR, Jiang D-H, Jackson V, Fahn S (1982) The effect of chronic L-DOPA administration on supersensitivity pre- and post-synaptic dopaminergic receptors in rat brain. Life Sci 31: 37–44

Reisine TD, Fields JZ, Yamamura HI (1977) Neurotransmitter receptor alterations in Parkinson's disease. Life Sci 21: 335–344

Rinne UK (1987) Early combination of bromocriptine and levodopa in the treatment of Parkinson's disease: a 5-year follow-up. Neurology 37: 826–828

Rinne UK, Lonnberg P, Koskinen V (1981) Dopamine receptors in the parkinsonian brain. J Neural Transm 51: 97–106

Sahakian B, Carlsson KR, De Girolami U, Bhawan J (1980) Functional and structural consequences of long-term dietary L-DOPA treatment in mice. Commun Psychopharmacol 4: 169–176

Seeman P (1987) Dopamine receptors in human brain diseases. In: Dopamine receptors. Alan R Liss, New York, pp 233–245

Smith RC, Strong JR, Hicks PB, Samorajski T (1979) Behavioural evidence for supersensitivity after chronic bromocriptine administration. Psychopharmacology 60: 241–246

Stern GM, Lees AJ (1983) Sustained bromocriptine therapy in 50 previously untreated patients with Parkinson's disease. Adv Neurol 37: 17–21

Suga M (1980) Effect of long-term L-DOPA administration on the dopaminergic and cholinergic (muscarinic) receptors of striatum in 6-hydroxydopamine lesioned rats. Life Sci 27: 877–882

Trendelenburg U (1966) Mechanisms of supersensitivity and subsensitivity to sympatho-mimetic amines. Pharmacol Rev 18: 629–640

Weiner WJ, Carvey P, Nausieda PA, Goetz CG, Klawans HL (1981) The effect of chronic levodopa on haloperidol-induced behavioural supersensitivity in the guinea-pig. Life Sci 28: 2173–2178

Wilner KD, Butler IJ, Seifert WE, Clement-Cormier YC (1980) Biochemical alterations of dopamine receptor responses following chronic levodopa therapy. Biochem Pharmacol 29: 701–706

Authors' address: Dr. P. Jenner, Department of Neurology, Institute of Psychiatry, De Crespigny Park, Denmark Hill, London SE5 8AF, U.K.

J Neural Transm (1988) [Suppl] 27: 177–183

Effect of chronic subcutaneous minipump infusion of lisuride upon locomotor activity of rats

H. Wachtel, K.-J. Rettig, and **P.-A. Löschmann**

Research Laboratories of Schering AG, Berlin and Bergkamen,
Federal Republic of Germany

Summary. Male Wistar rats were infused continuously for 14 days with lisuride 0.25 mg/kg/day or with vehicle via subcutaneously implanted osmotic minipumps. Locomotor activity was measured at 5 hours, 1, 7 and 14 days after implantation. Thereafter the minipumps were removed and 1, 7 and 21 days later the locomotor activity was recorded after a subcutaneous challenge dose of lisuride 0.1 mg/kg. In the course of continuous infusion the lisuride-treated rats showed a persistent stimulation of locomotor activity which remained almost constant throughout the whole period of exposure. At all intervals after removal of the minipumps lisuride challenge produced a less pronounced locomotor stimulation in lisuride-infused rats compared to vehicle-infused animals. This observation contrasts with the findings after chronic subcutaneous bolus treatment of rats with 0.25 mg/kg lisuride once daily for 29 days which resulted (1) in a progressive enhancement of the locomotor stimulatory effect and (2) in a longlasting hyperresponsiveness towards a subcutaneous challenge dose of lisuride 0.025 mg/kg. These results are discussed with respect to the advantage of the constant availability of lisuride at central dopamine receptors for the management of patients with advanced Parkinson's disease showing fluctuations in motor performance probably related to the kinetics of conventional oral therapy.

Introduction

The ergot derivative lisuride is a highly effective dopaminergic agonist with potent antiparkinsonian activity (Schachter et al., 1979; Lieberman et al., 1979; Gopinathan et al., 1981; Rinne, 1983). Due to the good water solubility of the compound, the development of an intravenous formulation was possible (Dorow et al., 1980). In parkinsonian patients suffering from severe "on/off dyskinesias" the continuous subcutaneous infusion of lisuride, in combination with oral L-dopa therapy, is very effective in reducing the motor fluctuations (Obeso et al., 1986; Stocchi et al., 1986).

In rats lisuride increased progressively the locomotor activity in the course of chronic subcutaneous (s.c.) treatment with 0.25 mg/kg once daily during 29 days (Wachtel et al., 1980). The enhanced response towards the locomotor stimulatory effect of an acute challenge dose of 0.025 mg/kg lisuride s.c. persisted up to 21 days following the termination of chronic lisuride treatment.

To test the hypothesis that the repeated bolus injection of lisuride might cause the hyperresponsiveness, lisuride was infused continuously via subcutaneously implanted osmotic minipumps during 14 days. Locomotor activity was recorded in the course of continuous infusion and after subcutaneous bolus challenge with lisuride following removal of the minipumps.

Material and methods

24 male Wistar rats weighting 100–120 g (Schering AG, Berlin) at the beginning of the experiment were used. The animals were kept under standard conditions in a temperature-controlled room (22° ± 1 °C) with a 12-hour light/dark cycle (light on from 6 a.m. to 6 p.m.). The animals received a standard diet (Altromin, Altromin Spezialfutterwerke GmbH, Lage, Federal Republic of Germany) and tap water ad libitum.

Drugs and solutions

For the continuous application osmotic minipumps (Alzet, Alza, Palo Alto, U.S.A.) with the following specifications were used: mean max. volume: 245.5 µl, mean max. pumprate: 0.49 µl/hour, mean max. pumping time: 500 hours. The minipumps were filled with their mean maximal volume of a solution containing 4.56 mg/ml lisuride hydrogen maleate (i.e. 3.4 mg/ml free base) dissolved in physiological saline solution containing 34 mg/ml tartaric acid. Controls were implanted with minipumps containing the corresponding volume of vehicle. The daily delivered dose of lisuride hydrogen maleate was 0.25 mg/kg/24 hours as based on a mean body weight of about 160 g. After removal of the minipumps the residual volume was evacuated, measured and frozen for later lisuride determination. The average recovery was more than 80%, indicating only minor decomposition of the compound in the course of continuous infusion.

For acute injections lisuride hydrogen maleate was dissolved in physiological saline solution, containing 1 mg/ml ascorbic acid.

Surgery

Under ether narcosis the minipumps were implanted subcutaneously on the back side. They were removed under ether narcosis after 14 days.

Measurements of locomotor activity

Locomotor activity of individual animals was measured using circular photocell activity cages similar to the method of Wright et al. (1962) and described in detail elsewhere (Wachtel, 1983). Briefly, the 12 activity cages consisted of two plastic cylinders forming a circular runway. The number of interruptions of six symmetrically placed light beams due to the animal's movements was determined and the counts accumulated for 10-minute intervals were recorded for 120 min.

Mean counts ± SEM were calculated for the two treatment groups. Statistical analysis of the differences between the means of the two groups was determined by the two-tailed t-test.

Experimental design

12 animals were implanted with osmotic minipumps filled with lisuride solution and another 12 animals with pumps containing vehicle. Of the lisuride group one animal developed an

abscess and was excluded from the study, another animal died during ether narcosis for removal of the minipumps. Locomotor activity was recorded on day 1 (5 hours after implantation), day 2 (24 hours after implantation), day 7 (167 hours after implantation) and on day 14 (333 hours after implantation). Immediately after the last run the minipumps were removed and prepared for later lisuride determinations. On day 15, 22 and 36 (i.e. 1, 7 and 21 days after removal) all animals were treated with 0.1 mg/kg lisuride s.c. and locomotor activity was recorded as described.

Results

The first measurement of locomotor activity 5 hours after implantation of the minipumps revealed no differences between lisuride- and control-treated animals (Fig. 1 A). At this time the minipumps start to deliver the solutions. On day 2,

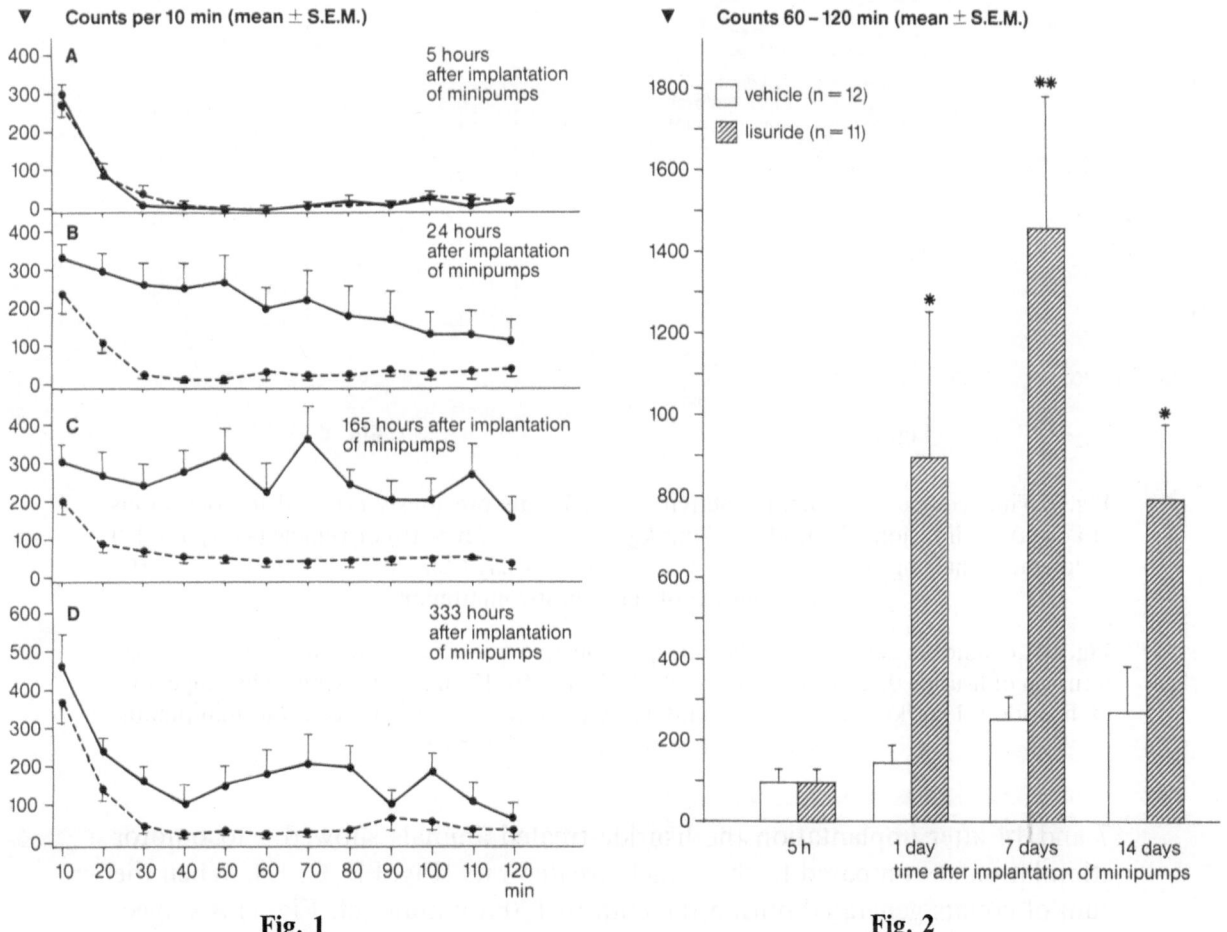

Fig. 1   **Fig. 2**

Fig. 1. Time course of locomotor activity of male rats during subcutaneous infusion of lisuride 0.25 mg/kg/day (————, n = 11) or vehicle (- - - -, n = 12) via osmotic minipumps 5 hours (**A**), 1 day (**B**), 7 days (**C**), and 14 days (**D**) after implantation of the minipumps

Fig. 2. Locomotor activity of male rats continuously infused with lisuride 0.25 mg/kg/day or vehicle during 60–120 min at various intervals following subcutaneous implantation of minipumps. Statistics: two-tailed t-test, *p < 0.05, **p < 0.01

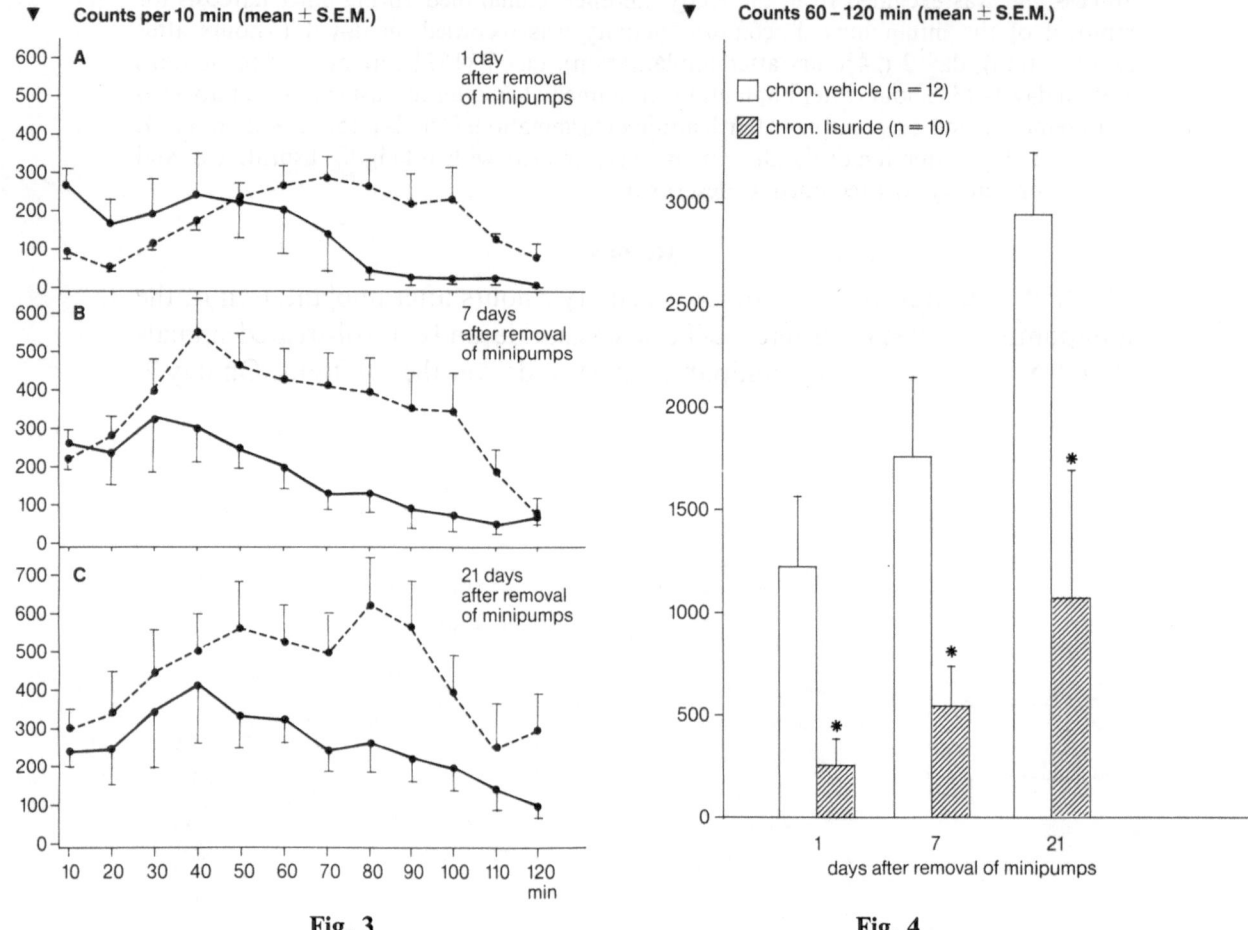

Fig. 3. Time course of locomotor activity of male rats previously exposed to continuous (14 days) s.c. infusion of lisuride 0.25 mg/kg/day (————, n = 10) or vehicle (- - - -, n = 12) following a challenge dose of lisuride 0.1 mg/kg s.c. 1 day (**A**), 7 days (**B**), and 21 days (**C**) after removal of the osmotic minipumps

Fig. 4. Locomotor activity of male rats previously exposed to continuous (14 days) s.c. infusion of lisuride 0.25 mg/kg/day or vehicle during 60–120 min following a challenge dose of lisuride 0.1 mg/kg s.c. at various intervals after removal of the osmotic minipumps. Statistics: two-tailed t-test, *p < 0.05

7 and 14 after implantation the lisuride-treated animals showed a locomotor stimulation as compared to the vehicle-treated rats (Fig. 1 B, C, D). When the sum of counts generated during the 60th to 120th minute (cf. Fig. 1) is subject to statistical analysis, the lisuride-infused rats exhibited a significant higher locomotor activity than vehicle-infused animals on day 2, 7 and 14 (Fig. 2). The extent of locomotor stimulation on day 7 and 14 practically did not differ from that at day 2 (see Figs. 1 and 2).

When challenged with 0.1 mg/kg lisuride s.c. on day 15, 22 and 36 (i.e. 1, 7 and 21 days after removal of minipumps) the vehicle- as well as the lisuride-exposed rats showed a clearcut locomotor stimulation (Fig. 3A, B, C). The extent and the duration of the locomotor stimulatory action of lisuride were attenuated in the rats continuously exposed to lisuride. This is further illustrated in Fig. 4 where the sums of counts during the second hour after lisuride challenge are depicted. In comparison with the vehicle-exposed rats the lisuride-treated animals showed a significantly less pronounced locomotor stimulation. During the first hour there was no statistically significant difference between both groups (data not shown). However, it should be pointed out that the initial locomotor depression seen in vehicle-pretreated rats after lisuride challenge on day 1 after minipump removal was attenuated in continuously lisuride-exposed rats (Fig. 3A). Interestingly there was a progressive increment in the extent of locomotor stimulation following the repeated subuctaneous bolus injection of lisuride in both vehicle- and lisuride-exposed animals even at 7 or 14 days challenge intervals (Fig. 3B, C).

The mean body weight of the lisuride-exposed animals did not differ from the vehicle-exposed animals (data not shown).

Discussion

In the course of continuous infusion the lisuride-treated rats showed a persistent stimulation of locomotor activity which remained almost constant throughout the whole period of exposure. At all intervals after removal of the minipumps lisuride challenge produced a less pronounced locomotor stimulation in lisuride-infused rats compared to vehicle-infused animals. In contrast to the previously published results after repeated subcutaneous bolus injection of lisuride 0.25 mg/kg for 14 days (Wachtel et al., 1980) no progressive increase of the locomotor activity of rats was seen in the course of continuous infusion of 0.25 mg/kg/day over the same period of time. The attenuation of the initial locomotor depression in the continuously lisuride-infused rats observed after a bolus challenge dose of lisuride 0.1 mg/kg s.c. on day 1 after minipump removal could reflect central dopamine (DA) autoreceptor subsensitivity in these animals due to permanent DA agonist exposure. The less pronounced locomotor stimulatory effect concerning extent and duration seen after bolus challenges with lisuride 0.1 mg/kg s.c. in the continuously lisuride-infused rats could reflect (1) a decreased susceptibility towards the DA receptor activating action of lisuride due to adaptive reactions following sustained agonist availability at central DA receptors or (2) a qualitative change in the motor response to the effect that stereotypies predominate to the exclusion of locomotor activation. The latter is unlikely in view of the observation that the extent of locomotor stimulation progressively increased after repeated daily bolus administration of lisuride for up to 29 days (Wachtel et al., 1980). In addition gross behavioral observation did not reveal stereotypies. Induction of lisuride-metabolizing enzymes also appears a less likely explanation for the above effect as the extent of the locomotor stimulation did not decline but remained almost unaltered during continuous infusion.

The increment in the locomotor stimulation in the course of repeated bolus challenges with lisuride 0.1 mg/kg s.c. in the vehicle – as well as the lisuride-infused rats lying as long as 7 and 14 days apart indicates that bolus administration of lisuride triggers some kind of hyperresponsiveness to the DA agonistic action of the compound which also might be of correlate a dyskinetic phenomena occasionally occurring in patients with Parkinson's disease treated with single oral doses of lisuride or other DA agonists. Our findings agree with previously published data obtained with other DA agonists showing that the absence of drug-free periods, in contrast to intermittent treatment, does not cause sensitization, but rather a decline, to the stereotyped behavior of amphetamine (Nielsen, 1981) or to the locomotor stimulation of apomorphine (Castro et al., 1985) during chronic drug treatment. Continuous infusion of lisuride with the result of sustained availability of the compound at central DA receptors seems to avoid the development of hyperresponsiveness and excessive stimulation of DA receptors as seen after repeated bolus administration of the drug and therefore might represent the more advantageous mode of DA agonist substitution therapy for patients with advanced Parkinson's disease (Quinn et al., 1982; Obeso et al., 1986; Stocchi et al., 1986; Stibe et al., 1987).

Acknowledgements

We thank Dr. Eck, Scherling AG, for lisuride determinations and U. Müller for technical assistance.

References

Castro R, Abreu P, Calzadilla CH, Rodriguez M (1985) Increased or decreased locomotor response in rats following repeated administration of apomorphine depends on dosage interval. Psychopharmacology 85: 333–339

Dorow R, Gräf KJ, Nieuweboer B, Horowski R (1980) Intravenous lisuride: a new tool for testing responsiveness to dopaminergic agonists and neuroendocrine function. Acta Endocrinol [Suppl 234] (Copenh) 94: 9

Gopinathan G, Teraväinen H, Dambrosia JN, Ward CD, Sanes JH, Stuart WK, Evarts EV, Calne DB (1981) Lisuride in parkinsonism. Neurology 31: 371–376

Lieberman A, Leibowitz M, Neophytides A, Kupersmith M, Mehl S, Kleinberg D, Serby M, Goldstein M (1979) Pergolide and lisuride for Parkinson's disease. Lancet 2: 1129–1130

Nielsen EB (1981) Rapid decline of stereotyped behavior in rats during constant one week administration of amphetamine via implanted Alzet osmotic minipumps. Pharmacol Biochem Behav 15: 161–165

Obeso JA, Luquin MR, Martinez-Lage JM (1986) Lisuride infusion pump: a device for the treatment of motor fluctuations in Parkinson's disease. Lancet 1: 467–470

Quinn N, Marsden CD, Parkes JD (1982) Complicated response fluctuations in Parkinson's disease: response to intravenous infusion of levodopa. Lancet 2: 412–415

Rinne UK (1983) New ergot derivatives in the treatment of Parkinson's disease. In: Calne DB, Horowski R, McDonald RJ, Wuttke W (eds) Lisuride and other dopamine agonists. Raven Press, New York, pp 431–442

Schachter M, Blackstock J, Dick JPR, George RJD, Marsden CD, Parkes JD (1979) Lisuride in Parkinson's disease. Lancet 2: 1129

Stibe C, Lees A, Stern G (1987) Subcutaneous infusion of apomorphine and lisuride in the treatment of parkinsonian on-off fluctuations. Lancet 1: 871

Stocchi F, Ruggieri S, Brughitta G, Agnoli A (1986) Problems in daily motor performances in Parkinson's disease: the continuous dopaminergic stimulation. J Neural Transm [Suppl] 22: 209–218

Wachtel H, Zehleke P, Schlangen M (1980) Supersensitivity following the chronic administration of the dopaminergic ergot derivative lisuride hydrogen maleate (LHM) to rats. Naunyn Schmiedebergs Arch Pharmacol [Suppl] 311: R 70

Wachtel H (1983) Central dopaminergic and antidopaminergic effects of ergot derivatives structurally related to lisuride. In: Calne DB, Horowski R, McDonald RJ, Wuttke W (eds) Lisuride and other dopamine agonists. Raven Press, New York, pp 109–125

Wright LS, Horn HJ, Woodard G (1962) Activity patterns in mice tested singly and in groups as a drug screening tool. Fed Proc 21: 420

Author's address: Dr. H. Wachtel, Department of Neuropsychopharmacology, Schering AG, Müllerstrasse 170–178, D-1000 Berlin 65.

J Neural Transm (1988) [Suppl] 27: 185–187

Discussion

Chronic infusion of levodopa and DA agonists in experimental models
of Parkinson's disease

Horowski: You have mentioned toxic effects of levodopa metabolites as a possibility. Would this apply to 3-O-methyldopa or would you consider other metabolites as well? I remember when one has solutions of levodopa or dopamine, they turn dark after some time and that is supposed to be the formation of quinones. Is there any evidence that quinones or even epoxides are being formed in vivo?

Melamed: Not to my knowledge, but when we just put levodopa into solution and hold it at room temperature in the light it turns dark. That is also one of the problems why we can't give levodopa by infusions for a long period of time. I don't know whether 3-O-methyldopa is the major problem.

Stahl: Have you looked further into the possibility that chronic dopa administration is decreasing the affinity of dopa decarboxylase, or decreasing the affinity for amino acid transport? That could explain why you don't see it with oral administration, but you do see it with intraperitoneal. Because if you have a huge peak which is a relatively pulsatile delivery and you have a system which has a decreased affinity, you will lose the pulses' ability to transfer that dopa into dopamine. It suggests to me that those would be very high candidates, a decreased affinity either for the decarboxylase or for the transport system. Have you looked at that?

Melamed: No, we haven't looked at that. People from Japan, I think in Tokyo, looked at dopa decarboxylase activity in the striatum following chronic administration of levodopa in vivo. They found that dopa decarboxylase in the striatum is indeed decreased following chronic administration.

With regard to the penetration to the blood-brain barrier, I also thought that maybe something happens to the transport apparatus following chronic bombardment by huge amounts of levodopa, but up to now I don't have any evidence for that.

Rinne: As you described in MPTP-monkeys, also in post-mortem brains of Parkinson patients treated with levodopa you can find an increase of dopamine level, but it is significantly less than in controls.

Leenders: You said exogenously administered L-dopa is mainly not stored as dopamine in the vesicular compartment. But isn't it so that the MPTP-slide you showed is evidence for that. The more the nerve terminals of the nigrostriatal pathway are lost, the less dopamine is there. How would you see that?

Melamed: I am not saying that dopamine which is formed from levodopa is not stored in any fashion in the striatum. What I am trying to say, and I think this is a major issue, is that the dopamine that is formed from exogenous levodopa is not stored in a vesicular form in the striatum, meaning that it would not be subject to demand release into the synapse. But it could be stored in other compartments, for instance it could be stored in the cytosol itself. I believe that at least in the remaining dopaminergic terminals there is cytosolic storage of dopamine.

Nutt: Although, like you, we are impressed that absorption is a major problem, I don't think that there is an alteration in the absorption of the drug with progression of the disease to account for the problems. Steve Gancher looked at the absorption in untreated patients, stable patients and fluctuating patients, at two different doses 1 and 3 mg/kg. If anything, the untreated patients absorbed the drug slightly less well than did other patients. I think erratic absorption clearly is a major problem in the patients that are very sensitive to the drug, but it is not something that develops with the progression of the disease, as far as we can tell.

Melamed: We did the same experiment. We looked at de novo patients and patients who were chronically treated with L-dopa but remained stable, and patients who were fluctuators. We monitored plasma levodopa levels not after infusion, but after their usual regular dose of levodopa in the morning and also at another dose taken in the evening or late afternoon, hoping to find differences in pharmacokinetics of levodopa in the fluctuators as opposed to the de novo patients and the stable parkinsonians. And we couldn't find any.

Receptor changes during chronic dopaminergic stimulation

Horowski: When you did your binding studies and saw some reduced binding of ^{3}H-spiroperidol with pergolide, did you wash it out. Pergolide of course has a very high affinity, that is the one you would most likely expect to displace?

Jenner: In those experiments, the animals are not withdrawn from treatment, so there will be drug in the tissue. The tissues are washed 3 times and we hope that will remove pergolide. But I think the important factor is that we didn't see any change in K_D, in the dissociation constant. I would have expected a change in K_D if drug had been present.

Gessa: Why were you surprised of the lack of parallelism of dopamine receptor changes and changes in sensitivity? This, I think, is an old rule, whenever you give chronic haloperidol, you have a supersensitivity that is measured by thousandfold, while changes in receptors are known to be at maximum 30–40%.

Jenner: I would completely agree.

Gessa: You have suggested that D 1 receptor stimulation maybe is not causing motor stimulation, but maybe is impairing the effect of D 2 stimulation. In order to support this suggestion you have shown SK&F results. However, SK&F again is a drug which should be studied further; maybe it is not a pure D 1 stimulant. If we had to conclude that D 1 stimulation is causing an opposite effect to D 2 stimulation, then using apomorphine or N-propylnorapomorphine, which are both D 1 and D 2 stimulants, we would have probably no effect, while actually they are much more potent.

Jenner: Again, I would agree. I think, SK&F 38393 may not represent what D 1 agonists as a whole may do. There is other evidence to suggest that other D 1 agonists will synergize with the D 2 agonists.

Biochemical basis of diagnostic and therapeutic problems

J Neural Transm (1988) [Suppl] 27: 191–199

Factors contributing to fluctuations of the dopaminergic nigro-striatal feedback system in Parkinson's disease

P. Riederer

Clinical Neurochemistry Section, Department of Psychiatry, University Clinic, Würzburg, Federal Republic of Germany

Summary. The experience that the supplementation of depleted dopamine in the nigro-striatal system of parkinsonian patients with L-dopa improves the clinical triad, akinesia, rigidity and tremor, mainly applies to long-term treatment in the early phase of Parkinson's disease. Complications in motor performance, like on-off response, wearing-off phenomena, peak-dose dyskinesia, biphasic dyskinesia, off-period dystonia and others, after more than 3 to 5 years following the onset of treatment indicate fluctuations in the dopaminergic feedback control system. It is suggested that these complications are due to progressive presynaptic degeneration and late changes in postsynaptic receptor amplification. However, as fluctuations are not imperative in all patients, an important additional aspect seems to be the topography of denervation, which involves different portions of the striatum to varying degrees. Location and extent of denervation are criteria which appear to have predictive value for the malignancy of the disease, the therapeutic response of drugs and complications in long-term treatment.

Open problems in clinical practice

The armamentarium for the treatment of idiopathic Parkinson's disease (PD) comprises a variety of antiparkinson drugs including L-dopa plus the peripherally acting decarboxylase inhibitors (DI) benserazide and carbidopa, amantadine, the monoamine oxidase (MAO)-B inhibitor (−)deprenyl, and dopaminergic (DA) agonists like bromocriptine and lisuride. Undoubtedly, the most extensive clinical experience has been gathered with the use of L-dopa plus DIs. Originally it was suggested that the supplementation of depleted dopamine (DA) by the precursor amino acid reasonably should improve akinetic symptoms. In fact, this assumption has proved to be valid only for long-term treatment in the early phase of PD, while complications in motor behavior are seen after prolonged treatment with L-dopa applied for more than 3–5 years (Marsden et al., 1982), including on-off phases and end-of-dose deterioration (wearing-off) in about 10–15% of patients.

Furthermore, L-dopa-induced dyskinesias already occur early in the course of the disease including choreatic athetosis of the extremities, trunk and head/face as well as various types of dystonias. These phenomena seem to correlate with the efficacy of L-dopa treatment as the clinical differentiation between peakdose dyskinesias, biphasic dyskinesias and off-period dystonia has been suggested (Marsden et al., 1982; Muenter et al., 1977; Birkmayer and Riederer, 1985). Although the pathophysiological mechanisms underlying these fluctuations are still unknown, several hypotheses have led to the assumption that disturbances in the feedback control of the nigro-striatal system are to blame (Carlsson, 1983).

Another clinical phenomenon still to be investigated is the diverging responsiveness of younger (< 50 years) and older (> 70 years) patients as well as benign and malignant courses of the disease to L-dopa therapy (Birkmayer et al., 1979). In addition, there seems to be a time lag in the kinetic response to L-dopa and in the onset of psychotic episodes, as compared to a more rapid kinetic response, as well as in the amelioration of psychosis and the occurrence of akinesia following L-dopa withdrawal.

Pathophysiological concepts

Presynaptic denervation

Circadian or long-term oscillations of motor performance are suggestive of disturbances in the feedback control of the nigro-striatal/strio-nigral loop. Under physiological conditions DA is synthetized from endogenous precursor amino acids, including dopa, and most efficiently stored in vesicles. Its release into the synaptic cleft is controlled by nerve impulses and regulated by pre- and postsynaptic receptor activities. In contrast, denervation of presynaptic neurons in PD progressively decreases the capacity of synthetizing DA from tyrosine and dopa. Reduced storage capacity and a decline in DA release are supposed to be the pathophysiological underpinnings of PD. In fact, supplementation of DA is able to overcome akinesia for some time, although there is no relationship between the decrement of presynaptic terminals and the responsiveness to L-dopa. A wide individual margin in the sensitivity of behavioral response (akinesia) to L-dopa (Fig. 1) is indicative of a disturbed nigro-striatal DA feedback control. As long as this feedback control can be put into adequate operation, the incidence of akinesia can be minimized.

Compensatory mechanisms of endogenous origin, denoted by a decrease in the DA/homovanillic acid (HVA) ratio, aim to enhance the release of DA at a stage at which denervation has already been accomplished. If the loss of neurons exceeds a threshold of about 70%, clinical symptoms become apparent and can be improved by replenishing missing DA. However, the exogenously administered precursor L-dopa is applied in unphysiological quantities most of the time. Depending on the remaining vesicular storage capacity, more or less

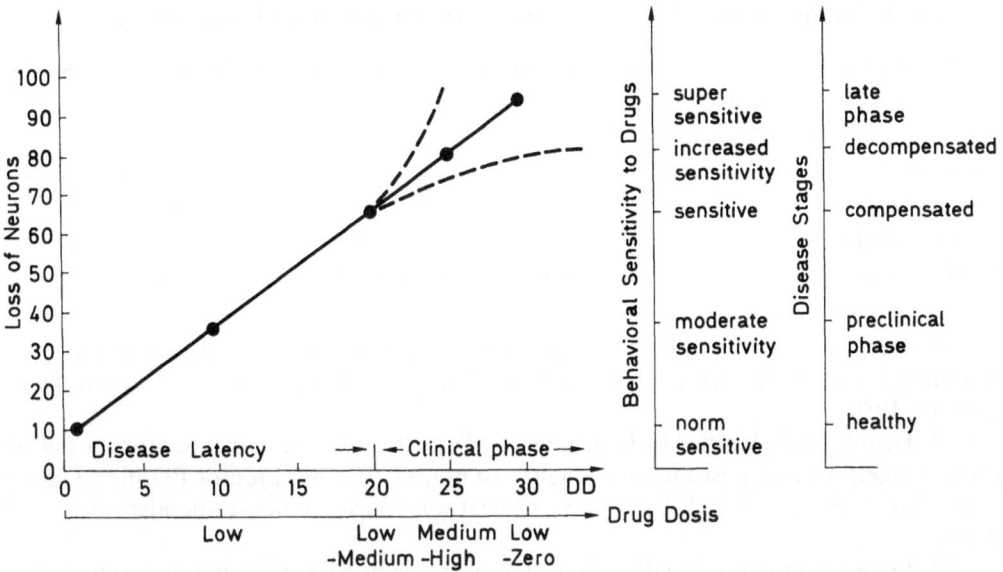

Fig. 1. A hypothetical representation correlating denervation of the nigro-striatal dopaminergic system in Parkinson's disease (PD) with behavioral sensitivity to drugs, their dose and stages of the disease. According to various concepts, denervation might not be a linear process but may accelerate along with the duration of PD. *DD* Duration of the disease

DA, synthetized from a small pool of endogenous and a large pool of exogenous L-dopa, will be stored and released using the nerve impulse traffic as a mediator.

As the storage capacity of vesicles declines in PD with progressive denervation, growing amounts of DA derived from exogenously administered L-dopa will not be utilized for receptor stimulation via physiological routes. Additional uncontrolled leakage into the synaptic cleft may occur and DA, synthetized in other compartments, may then contribute to synaptic processes. This kind of uncontrolled receptor stimulation presumably impairs the endogenous kybernetic system maintaining feedback control. Even an unphysiological stimulation for presynaptic receptors would be without consequences for regulator processes as the generation of DA via tyrosine hydroxylase would be minimized accordingly, but the effects induced by excessive amounts of exogenously applied L-dopa in the treatment of PD cannot be counterbalanced any more.

Postsynaptic dysfunction

The unphysiological stimulation of postsynaptic receptors induces their downregulation which might end up in a hyperpolarization block of the receptors due to excessive DA. Postsynaptic DA receptors are distinguished by two subclasses, adenylate cyclase-dependent DA receptors and D2 receptors independent from this enzymatic process (Kebabian and Calne, 1979). In PD, these receptors display decreased, unchanged or increased densities as defined by antagonist binding techniques (Birkmayer and Riederer, 1985). These discrep-

Table 1. ^{3}H-Spiroperidol binding in post-mortem putamen of Parkinson's disease*

Parkinson's disease treated with L-dopa + DI, amantadine sulfate**	B_{Max} pmol/g	K_D (nM)
Responder [15]	19.3 ± 1.42	0.18 ± 0.03
Non-responder [5]	19.8 ± 1.55	0.20 ± 0.021
Controls [25]	20.6 ± 1.87	0.14 ± 0.027

Data as mean ± sem; number of patients in parenthesis; B_{Max} and K_D values were determined from Scatchard analysis; receptor binding technique was done according to Seeman, 1980

  * Histologically verified by K. Jellinger (1986); all groups were matched for age (69–80 years; range), sex and post-mortem time (< 10 hours). The duration of PD in the group responding to drugs was 11.5 ± 2.3 years, while it was shorter in non-responders (8.7 ± 2.1 years)

  ** Long-term treatment with L-dopa plus benserazide 62.5–187.5 mg/day; amantadine sulfate 300–400 mg/day for responder and non-responder

ancies, however, are mostly due to different drug regimens employed (Guttman et al., 1986). As a result of the largest study, Seeman et al. (1987) report on a small but significant increase in D 2 receptor density in untreated patients and no change in a treated subject group (Guttman et al., 1986). These techniques, however, only provide information as to which degree ligands bind to a receptor protein, whereas no evidence can be elicited from such data regarding the quality of the amplification- and output system.

At least in long-term treatment and in the decompensated late phase of PD malfunction of postsynaptic receptors may be conspicuous. This is evident in about 30% of patients with PD not responding to antiparkinson drugs any longer and dying in an akinetic crisis but exhibiting almost normal D 2 receptor densities post mortem (Table 1). In such cases, earlier studies have demonstrated an impaired stimulus of adenylate cyclase towards DA (Riederer et al., 1978). In addition, displacement of radiolabeled spiroperidol in post-mortem putamen by classical DA agonists, like bromocriptine and lisuride, suggests a tendency towards increased receptor sensitivity (lower IC_{50} values), whereas a decline was noteable for compound CI 201-678 on the other hand. The mixed agonist/antagonist terguride did not alter these properties (Table 2). Although these preliminary data may only be indicative, opposite shifts induced by the D 1 and D 2 agonist CI 201-678 and the D 2 agonists bromocriptine and lisuride should further be examined in denervated tissue.

These data support evidence to presume that postsynaptic receptor lose their functional acitivity in late phases of PD. Noteworthy in this context is the experience that amantadine, able to change membrane fluidity (Wesemann, 1984) by a mechanism yet unknown, apparently is a valuable drug in the management of akinetic crises (Danielczyk, 1973). The growing rigidity of

Table 2. Displacement of ^{3}H-spiroperidol binding (870 pM) in human putamen by dopamine agonists

	Control IC$_{50}$ (nM)	Parkinson's disease (IC$_{50}$ (nM)
CI 201-678	300	350
Bromocriptine	38	30
Lisuride	9	7
Terguride	11	11

neuronal membranes might disturb receptor integrity by blocking the conformational change between low and high affinity states. Such conformational changes, physiologically induced by the endogenous ligand DA, seem to be of significance for the pharmacological effectiveness of DA agonists. Goldstein et al. (1985) hypothesize that DA agonists with low intrinsic activity require the presence of synaptic DA (therapeutically formed by L-dopa) in order to be able to mediate an agonistic effect.

Topographic aspects of denervation

Denervation of the substantia nigra (SN) does not show a uniform and unspecific pattern but is characterized by a unilateral or, more frequently, symmetrical focal neuronal loss, mainly affecting the central and caudal parts of the zona compacta of the SN (Jellinger, 1986). Therefore, striatal denervation may be defined topographically. This approach is substantiated by the arrangement of the nigro-striatal projection, innervating the putamen mainly from the caudolateral portion of SN zona compacta, while the caudate nucleus draws its DA fibres from the rostromedial SN zona compacta. More recently it has been demonstrated that the subregional loss of DA in the putamen and the caudate nucleus even present opposite patterns. The rostral part of the caudate head is more severely affected than the caudal part (downturn to 16% vs. 38%), while the rostral part of the putamen (13%) is less damaged than the caudal part (4%; Hornykiewicz and Kish, 1986). Our own data comply with this finding, proving a decrease of DA down to 11% in the caudal part of the putamen, while the rostral part is less affected (39%; Riederer et al., 1986). In addition, the more severely denervated caudal putamen gives evidence of a more severe downturn of DOPAC (19%) and HVA (41%), while these disturbances are less marked in the rostral part (DOPAC 79%, HVA 72%, Table 3).

Table 3 also provides evidence of the predominantly presynaptic localization of DOPAC. DOPAC is a more sensitive marker for the denervation process than HVA. In the more severely affected caudal putamen the reduction of DOPAC matches that of DA while the loss of DOPAC approximates that of HVA in the less affected portion (Riederer et al., 1986). The concordance of

Table 3. Dopamine, DOPAC, and HVA in the caudal and rostral putamen of benign Parkinson's disease (n = 3–5)

Putamen	Dopamine	DOPAC	HVA
	mean as percent of 3–4 controls		
Caudal	11	19	41
Rostral	39	79	72

	N	Age	Sex	Duration of disease (years)	Post-mortem time (hours)	Therapy	Diagnosis	Neuropathology
Controls	4	77 (74–83)	1M/3F		7.5 (4–11)	antibiotics cardiovascular treatment	heart infarction (2), bronchopneumonia (2)	age related brain histology
M. Parkinson	5	79 (75–82)	2M/3F	10 (8–13)	6.5 (3–9)	combined L-dopa amantadine	Parkinson's disease (5)	Lewy bodies loss of melanin in SN

From Riederer et al., 1986

DA concentrations in the rostral and caudal putamen is paralleled by D2 receptor densities in controls, as measured by antagonist bindings with ^{3}H-spiroperidol: In the caudal part the B_{max} reaches $19,4 \pm 5,1$ (5) pmol/g tissue (mean $\pm$ SEM) and K_D $0,35 \pm 0,04$ (5) nM. Similarly, in the rostral part of the putamen the B_{max} arrives at $19,3 \pm 5,3$ (5) pmol/g tissue and the K_D at $0,56 \pm 0,12$ nM.

The suggestion is put forward that topographic denervation of SN affects different areas of the striatum to a varying extent, and presumably receptors will become supersensitive only in subareas where the loss of innervation reaches a level of more than 90%. According to histological studies (Jellinger, 1986), however, such a high threshold of degeneration (if at all) apparently is reached in a few striatal areas only. In these parts the percentage increase of ^{3}H-spiroperidol binding, however, might exceed the elevation mark of 129% described by Seeman et al. (1987).

In fact, most of these data comply with the upper range of values measured in the control population. This again may be taken as evidence of a few circumscribed areas of supersensitivity, whereas most of the receptors are innervated, at least in part. Assuming that this hypothesis is valid, drug therapy would have to meet the demands of norm-sensitive, activated as well as supersensitive receptors. As a consequence, the physiological and behavioral output of therapeutic strategies would largely depend on extent and topography of denervation. The therapeutic effectiveness as well as side effects related to the motor system may be defined according to the complexity of the lesions, especially with regard to the subareas involved.

Therapeutic strategies to overcome feedback failure and motor fluctuation

Numerous attempts have been made to compensate fluctuations in the dopaminergic nigro-striatal feedback system by employing various therapeutic strategies. These include (1) improved galenical formulations of L-dopa, e.g. a sustained-release tablet, depot, slow and fast forms, infusion techniques (Hardie et al., 1986; Poewe et al., 1986; Kapfhammer et al., 1985), (2) new drugs to influence auto-, D1 and D2 receptors (Karobath, 1986; Goldstein et al., 1985; Clark et al., 1985; Corsini et al., 1984; Brücke et al., 1985), (3) lisuride infusions to maintain constant D2 receptor stimulation (see publications in this volume), (4) combinations of a variety of drugs with different pharmacological profiles, such as L-dopa + DI, the MAO-B inhibitor ($-$)deprenyl, amantadine and DA agonists (Birkmayer and Riederer, 1985; Rinne, 1986).

All these therapeutic maneuvres are aimed to maintain at least a certain level of dopaminergic function guaranting sufficient feedback regulation of intra- and interneuronal origin. However, on account of the progressive course of denervation and its topographic variation, feedback regulation will differ enormously in striatal subareas. This aspect, in particular, might determine malignancy and therapeutic effectiveness in PD.

References

Birkmayer W, Riederer P, Youdim MBH (1979) Distinction between benign and malignant type of Parkinson's disease. Clin Neurol Neurosurg 158: 81–83

Birkmayer W, Riederer P (1985) Die Parkinson-Krankheit, 2nd edn. Springer, Wien New York

Brücke T, Danielczyk W, Simanyi M, Sofic E, Riederer P (1987) Terguride: partial dopamine agonist in the treatment of Parkinson's disease. In: Yahr MD, Bergmann KJ (eds) Advances in neurology. Raven Press, New York, pp 573–576

Carlsson A (1983) Are "On-off" effects during chronic L-dopa treatment due to faulty feedback control of the nigrostriatal dopamine pathway? J Neural Transm [Suppl] 19: 153–161

Clark D, Hjorth S, Carlsson A (1985) Dopamine receptor agonists: mechanisms underlying autoreceptor selectivity. J Neural Transm 62: 171–207

Corsini GU, Horowski R, Reiner E, del Zompo M (1984) Treatment of Parkinson's disease with a dopamine partial agonist. Clin Neuropharmacol 7: 950–951

Danielczyk W (1973) Die Behandlung von akinetischen Krisen. Med Welt 24: 1278–1282

Goldstein M, Lieberman A, Meller E (1985) A possible molecular mechanism for the antiparkinsonian action of bromocriptine in combination with levodopa. TIPS (Nov.) 436–437

Guttman M, Seeman P, Reynolds GP, Riederer P, Jellinger K, Tourtellotte WW (1986) Dopamine D 2 receptor density remains constant in treated Parkinson's disease. Ann Neurol 19: 487–492

Hardie RJ, Malcolm SL, Lees AJ, Stern GM, Allen JG (1986) The pharmacokinetics of intravenous and oral levodopa in patients with Parkinson's disease who exhibit on-off fluctuations. Br J Clin Pharmacol 22: 429–436

Hornykiewicz O, Kish SJ (1986) Biochemical pathophysiology of Parkinson's disease. In: Yahr MD, Bergmann KJ (eds) Advances in neurology. Raven Press, New York, pp 19–34

Jellinger K (1986) Overview of morphological changes in Parkinson's disease. In: Yahr MD, Bergmann KJ (eds) Advances in neurology. Raven Press, New York, pp 1–18

Kapfhammer HP, Kuss HJ, Eben E, Rüther E (1985) L-dopa infusion experiments in combination with the L-dopa-decarboxylase inhibitors benserazide and carbidopa: L-dopa metabolism and neuroendocrinological effects. Pharmacopsychiat 18: 143–144

Karobath M (1985) Neue Aspekte in der Pharmakotherapie des M. Parkinson. In: Schnaberth G, Auff E (eds) Das Parkinson-Syndrom. Proceedings, Symposium 1984. Wissenschaftlicher Dienst „Roche", Basle, pp 211–216

Kebabian JW, Calne DB (1979) Multiple receptors for dopamine. Nature 277: 93–96

Marsden CD, Parkes JD, Quinn N (1982) Fluctuations of disability in Parkinson's disease: clinical aspects. In: Marsden CD, Fahn S (eds) Movement disorders. Butterworth, London, pp 86–122

Muenter MD, Sharpless NS, Tyce GM, Darley L (1977) Patterns of dystonia ("I-D-I" and "D-I-D") in response to L-dopa therapy of Parkinson's disease. Mayo Clin Proc 52: 163–174

Poewe WH, Lees AJ, Stern GM (1986) Treatment of motor fluctuations in Parkinson's disease with an oral sustained-release preparation of L-dopa: clinical and pharmacokinetic observations. Clin Neuropharmacol 9: 430–439

Riederer P, Rausch WD, Birkmayer W, Jellinger K, Seeman D (1978) CNS modulation of adrenal tyrosine hydroxylase in Parkinson's disease and metabolic encephalopathies. J Neural Transm [Suppl] 14: 121–132

Riederer P (1986) Neurobiochemische Aspekte zur Progression der Parkinson-Krankheit: Post-mortem Befunde und MPTP-Modell. In: Fischer PA (ed) Spätsyndrome der Parkinson-Krankheit. Proceedings, Parkinson-Symposium 1986, Frankfurt. Wissenschaft-licher Dienst „Roche", Basle, pp 37–48

Rinne UK (1986) The importance of an early combination of a dopamine agonist and levodopa in the treatment of Parkinson's disease. In: Van Manen J, Rinne UK (eds) Lisuride: a new dopamine agonist and Parkinson's disease. Proceedings of a symposium, Berlin 1985. Excerpta Medica, Amsterdam, pp 64–71

Seeman P (1980) Brain dopamine receptors. Pharmacol Rev 32: 229–313

Seeman P, Guan HC, Bergeron C, Reynolds GP, Bird ED, Riederer P, Jellinger K, Tour-tellotte WW (1987) Human brain D 1 and D 2 dopamine receptors in schizophrenia, Alzheimer's, Parkinson's, and Huntington's diseases. Neuropsychopharmacology, in press

Wesemann W (1984) Aspekte zum Wirkmechanismus von Amantadin. In: Danielczyk W, Wesemann W (eds) Amantadin-Workshop. Edition Materia Medica, Socio-Medico Verlag, Gräfelfing

Author's address: Dr. P. Riederer, Clinical Neurochemistry Section, Department of Psychiatry, University Clinic, Füchsleinstrasse 15, D-8700 Würzburg, Federal Republic of Germany.

J Neural Transm (1988) [Suppl] 27: 201–210

Agonist and antagonist actions of lisuride on dopamine neurons: electrophysiological evidence

G. L. Gessa

Department of Neurosciences, University of Cagliari, Cagliari, Italy

Summary. The effect of lisuride (LIS) on the firing rate of A 9 dopamine (DA) neurons in chloral-hydrate anesthetized and unanesthetized (paralyzed) rats was compared. In both preparations, the microiontophoretic application of LIS onto DA cell bodies consistently inhibited the electrical activity of the neurons. On the other hand, the effect of intravenous LIS differed in the two preparations. In anesthetized rats LIS (10–100 µg/kg) inhibited in a dose-related manner the firing rate of most DA neurons tested, whereas in unanesthetized rats LIS produced a dose-related increase in firing rate. The latter effect was transient, subsiding within 5 min, and was followed by the return of firing rate to baseline or slightly below it. Irrespective of the animal preparation, after the initial effect of LIS had subsided, DA neurons became totally insensitive to additional doses of LIS, to apomorphine and haloperidol. Such an insensitivity of DA neurons was present 1 to 6 but not 24 h after LIS (0.2 mg/kg, subcutaneously) treatment. A hypothesis is proposed to explain the different effects of LIS.

Introduction

The ergot derivative lisuride (LIS) is considered a direct dopamine (DA) agonist with a preferential action on DA autoreceptors (Seeman, 1981; Stoof and Kebabian, 1984; Tissari and Gessa, 1983).

Accordingly, while relatively high doses of LIS increase motor activity and produce stereotyped behavior (Horowski, 1978; Krejci et al., 1985), minute doses of the drug decrease locomotor activity, inhibit DA synthesis and DA striatal binding (for review, see Calne et al., 1983) and depress the electrical activity of DA neurons both in the Pars Compacta of the Substantia Nigra (A 9) and in the Ventral Tegmental Area (A 10) (Mereu et al., 1986 a; Walters et al., 1979, 1986; White and Wang, 1983). Although such effects mimic those of other DA agonists (Mereu et al., 1986 b; Skirboll et al., 1979), some data indicate that the action of LIS on DA neurons is more complex than that of a "pure" DAergic agonist. For instance, in chloral-hydrate (CHH-)anesthetized rats, a number of A 9 neurons were found to be only partially inhibited and a few even activated by LIS (Rogawski and Aghajanian, 1979). Moreover, following the partial suppression of the activity of A 9 or A 10 cells by LIS, additional doses of the drug or apomorphine (APO) were found to be unable to further depress the firing rate of these cells (Walters et al., 1979; White and Wang, 1983).

Finally following LIS or the other ergot derivative bromocriptine, haloperidol was found to be ineffective in reversing the firing depression induced by this drug (Bannon et al., 1980).

These peculiar effects of LIS have been interpreted in terms of the drug capability to act on DA autoreceptors as an "irreversible" agonist rather than somehow as a DA antagonist. The latter possibility was indeed ruled out because LIS was found unable to reverse the rate-depressant effect produced by APO on DA cells (Bannon et al., 1980; White and Wang, 1983).

However since the above electrophysiological studies have been carried out with anesthetized rats, and since general anesthetics have been shown to modify the response of DA neurons to different drugs (Gessa et al., 1987; Mereu et al., 1984) including ergot derivatives (Gessa et al., 1986 b; Mereu et al., 1984), it was of interest to compare the effect of lisuride on DA neurons in anesthetized and unanesthetized rats.

Methods

Male Sprague-Dawley rats were handled and scarified in strict accordance with the NIH "Guide for the Care and Use of Laboratory Animals" (DHEW Publication No. 80-23).

Two different preparations were used as already described in details (Mereu, 1982; Mereu et al., 1984, 1986 a, 1986 b, 1986 c).

Local anesthesia

These animals were anesthetized with halothane, tracheotomized and the trachea intubated with a catheter connected to a vaporizer through which a mixture of 2.0–2.5% halothane in air was delivered. A femoral vein was cannulated for injections and rats were restrained in a David Kopf stereotaxic apparatus. The wound edge and pressure points were largely infiltrated with a long-acting local anesthetic (2% mepivocaine hydrochloride) and diazepam (0.250 mg/kg) was administered intravenously (i.v.) in order to reduce the stress of the animal. D-tubocurarine (5.0 mg/kg) was injected i.v., the halothane discontinued and the tracheal catheter connected to an artificial respirator. During the experiments, additional doses of the local anesthetic and paralyzing agent were administered as needed.

General anesthesia

These rats were anesthetized with CHH (400 mg/kg i.p.) prepared as above, restrained in the stereotaxic apparatus and then paralyzed.

In order to obtain a steady state level of deep anesthesia, a solution of CHH (2% w/ v in 0.9% saline) was continuously infused i.v. at the rate flow of 1.5 ml/h (100 mg/kg/h).

The spontaneous extracellular action potentials from A 9 and A 10 DA cells were recorded and processed following conventional methodology (Bannon et al., 1980; Chiodo et al., 1984; Mereu, 1982; Mereu et al., 1984, 1986 a, 1986 b, 1986 c; Walters et al., 1979, 1986; White and Wang, 1983). DA cells were recognized on the basis of their well established electrophysiological, pharmacological and anatomical characteristics (Bannon et al., 1980; Chiodo and Bunney, 1984; Mereu, 1982; Mereu et al., 1984, 1986 a, 1986 b, 1986 c). For systemic administration lisuride hydrogen maleate and APO were dissolved in 0.9% saline.

Haloperidol and (−)-sulpiride were used as commercially available solutions. Drugs were administered i.v. in a fixed volume of 1 ml/kg via a the femoral cannula previously inserted.

For the microiontophoretic local application of the drugs, five-barrel micropipettes were used. The central recording barrel was filled with an electrolite saturated with 2% Pontamine Sky Blue (optimal impedance in vitro was 3–6 M Ω at 60 Hz in saline), while the outer barrels were filled as follows: 2 M NaCl for automatic current balancing; DA-HCl (0.1 M and 0.01 M, pH 4.5); LIS (0.01 M, pH 4.0); GABA (0.01 M, pH 4.5). The drugs were ejected as cations and retained with a current of − 10 nA.

Since no difference was observed between the effects of LIS on the activity of A 9 and A 10 neurons, data were pooled and the following results generally refer to DA neurons.

Results

Microiontophoretic application of LIS

The effect of iontophoretically applied LIS was tested on DA neurons in 4 CHH anesthetized and 4 unanesthetized rats. In each rat 2 to 3 cells were tested. The results in the two preparations were similar; namely LIS inhibited the firing rate, the degree of inhibition being directly proportional to the ejection current and inversely related to the basal firing rate of the cell (Mereu et al., 1986 c; Walters et al., 1979).

Moreover iontophoretically applied LIS failed to modify the sensitivity of the neuron to a subsequent application of DA (Fig. 1).

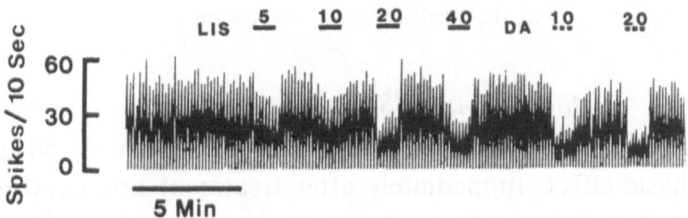

Fig. 1. Rate histogram showing the effect of microiontophoretically applied LIS and DA on the firing rate of a DAergic neuron. Horizontal bars indicate the length of current ejection; the number its intensity (mA)

Intravenous LIS in anesthetized rats

In agreement with previous reports (Bannon et al., 1980; Chiodo and Bunney, 1984; Gessa et al., 1986 b; Rogawski and Aghajanian, 1979; Walters et al., 1979, 1986; White and Wang, 1983), the intravenous injection of LIS in CHH anesthetized rats inhibited the firing rate of 16 out of 17 DA cells tested; the inhibition being dose-related (Fig. 2). Following a dose of 50 μg/kg the firing of DA neurons was almost suppressed and the recovery to baseline occurred after a mean period of 45 ± 7.3 min (Fig. 3 a). Some cells (N = 4) had not recovered to baseline more than one hour after the LIS injection.

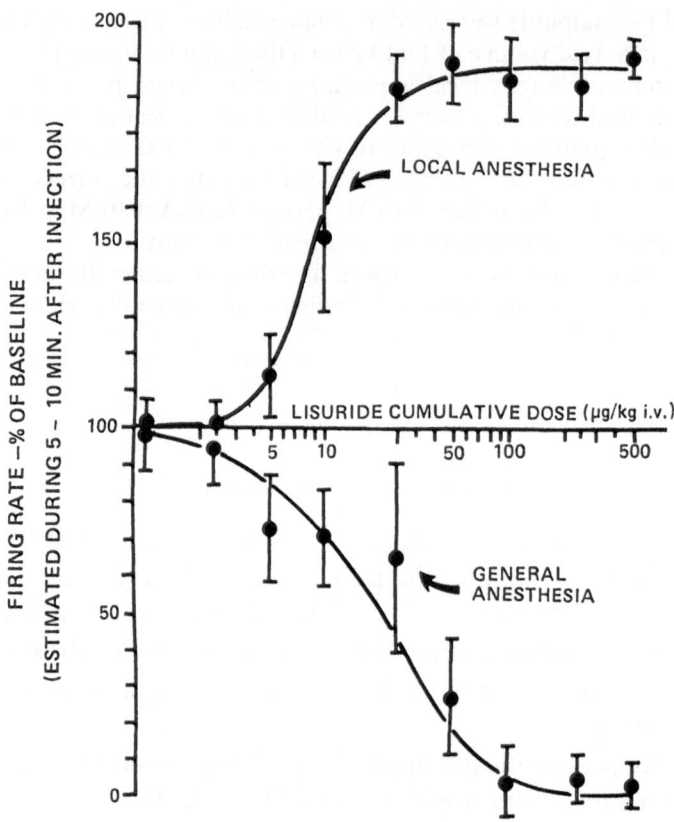

Fig. 2. Dose-curves of LIS effect on the firing rate of DA neurons in unanesthetized and chloralhydrate anesthetized rats. Values were monitored at 5–10 min after LIS treatment at the peak of maximal response

Intravenous LIS in unasthetized rats

As shown in Fig. 3 b in unanesthetized rats the injection of 50 µg/kg of LIS caused a biphasic effect. Immediately after treatment DA neurons showed a marked increase in firing rate (a mean of $66 \pm 9\%$ above baseline) which persisted for 6 to 10 min and then gradually subsided to baseline (8 cells) or to about 50% below it (6 cells). The stimulant effect of LIS was dose-related from 5 to 50 µg/kg (Fig. 2).

LIS-induced insensitivity of DA autoreceptors

Both in the anesthetized and unanesthetized preparations after the initial effect of LIS had subsided, additional doses of the drug (up to the cumulative dose of 200 µg/kg) were completely ineffective in modifying the firing rate.

Moreover, DA neurons became insensitive also to the inhibitory effect of APO (at the dose up to 200 µg/kg) and to the stimulant effect of sulpiride (100 mg/kg) and haloperidol (500 µg/kg) (see Fig. 3 c).

The time-course of LIS-induced "freezing" of DA neurons was studied. Rats were treated subcutaneously with 200 µg/kg of LIS, then, 1, 2, 6, and 24 h after treatment, they were prepared for single unit recording under CHH anesthesia.

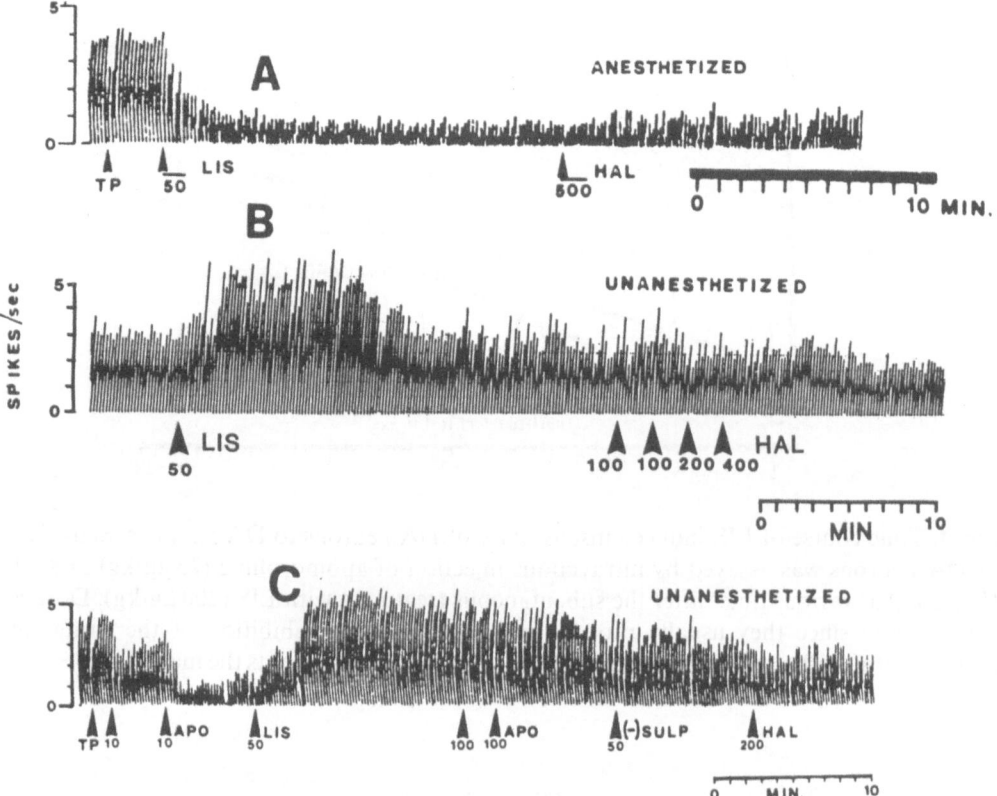

Fig. 3. Examples of the opposite effect elicited by systemic lisuride (*LIS*) on the firing rate of A 9 DAergic neurons in a CHH-anesthetized (**A**) and d-tubocurarine-paralyzed (**B**) rat. In the histogram **C**, the reversal by LIS of the apomorphine (*APO-*)induced inhibition, in a unanesthetized rat, is shown. Irrespective of the preparation, 10 min to 1 hour after the injection of a single bolus (50–100 µg/kg) of LIS, neither additional doses of the drug, nor APO, up to a cummulative dose of 500 µg/kg, were able to induce any effect. High doses of DA antagonists, such as (−)Sulpiride (*SULP*) or Haloberidol (*HAL*) administered i.v. during the depressory phase induced by LIS were unable to alter the firing rate of the cells

At 1, 2, and 6 h after LIS treatment DA neurons were almost totally insensitive both to APO and to LIS at the doses up to 200 µg/kg (Fig. 4). At 24 h the sensitivity of DA neurons to APO and to LIS had returned to the normal.

Apomorphine reversal

In both animal preparations intravenous APO (25 µg/kg) inhibited the firing rate of DA neurons (Mereu et al., 1986c; Rogawski and Aghajanian, 1976). The subsequent intravenous administration of LIS (50–100 µg/kg) produced no effect in the anesthetized preparation, whereas in unanesthetized rats it caused a rapid return of the firing rate to baseline (n = 4) or above it (n = 6) (Fig. 3 c).

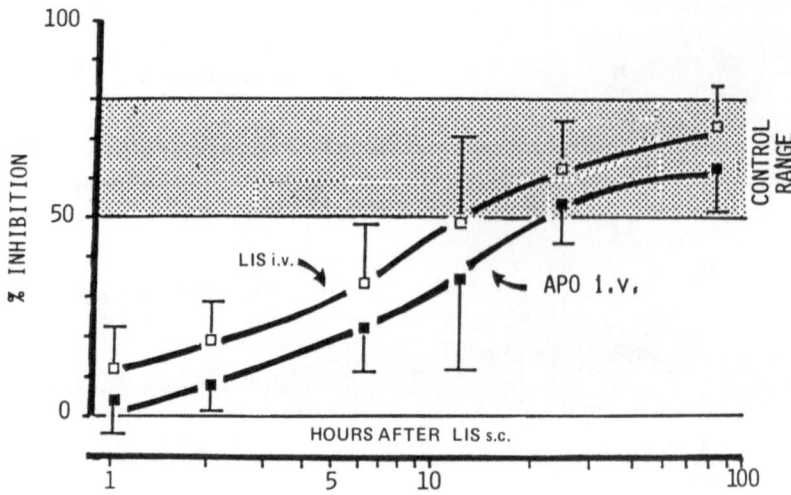

Fig. 4. Time course of LIS-induced insensitivity of DA neurons to DA agonists. Sensitivity of DA neurons was assayed by intravenous injection of apomorphine (25 μg/kg) and LIS (50 μg/kg) at various times after the subcutaneous treatment with LIS (200 μg/kg). Dosages were chosen since they usually produce in control rats an inhibition of the firing rate between 50 and 80% of baseline (dashed band). Each point represents the mean value ± S.E. obtained from 5 to 11 cells

Discussion

The results have shown that the iontophoretic application of LIS on DA cells consistently inhibited the firing of DA neurons, irrespectively of the animal preparation used.

On the other hand, the initial effect of systemically injected LIS varied depending on the experimental conditions. Namely, in agreement with previous observations (Bannon et al., 1980; Chiochio and Bunney, 1984; Gessa et al., 1986 b; Rogawski and Aghajanian, 1979; Walters et al., 1979, 1986; White and Wang, 1983), LIS inhibited the firing rate of most DA neurons if the experiments were carried out in generally anesthetized rats, but produced a transient stimulatory effect when the unanesthetized preparation was used. Moreover, in unanesthetized rats LIS readily reversed the APO-induced inhibition of firing, whereas it was ineffective in the anesthetized preparation.

Another outcome of our study is that LIS produced, both in anesthetized and unanesthetized rats, a delayed long lasting loss of the response of DA neurons to the effect of DA agonists (APO and LIS itself) and antagonists, such as haloperidol or sulpiride. Insensitivity of DA neurons to these agents was present soon after the recovery from the initial changes in firing and persisted for many hours.

These observations raise the question of the mechanism underlying the different electrophysiological changes produced by LIS and their relevance in the pharmacological actions of the drug.

In spite of the difficulties in interpreting single unit changes as an indication of the site and mode of action of a systemically injected drug, a possible interpretation for the contrasting effects of LIS might be proposed. According to our hypothesis LIS is a partial agonist for DA receptors having higher affinity but weaker intrinsic activity than the endogenous transmitter.

Accordingly, LIS would behave as agonist or antagonist depending on the ongoing state of activation of the DA receptor with which it interacts. Namely, LIS would act as antagonist when DA receptors are heavily impinged upon by endogenous DA, viceversa it would act as agonist when the receptor is functionally denervated. While postsynaptic DA receptors would fall into the first condition, there is evidence that DA autoreceptors are not tonically activated by DA (Chiodo and Bunney, 1984; Gessa et al., 1986 b; Llinas et al., 1984; Mereu, 1982). A similar hypothesis has been proposed by Carlsson for explaining the mechanism of action of the putative DA autoreceptor agonist 3-PPP (Clark et al., 1985).

In agreement to our hypothesis, the fact that LIS behaves a DA agonist inhibiting the firing rate of DA neurons when microiontophoretically applied to the somato dendritic region is expected. On the other hand the finding that systemically injected LIS acts as a DA agonist in the anesthetized preparation might be explained with the fact that general anesthesia impairs the striato-nigral feedback loop (Gessa et al., 1986 a, 1986 b; Mereu et al., 1984) and causes a sort of chemical, reversible, deafferentation of DA cells. Therefore, it is likely that in the anesthetized preparation any effect of LIS of postsynaptic DA receptors cannot be translated back to DA neurons, so that LIS may affect DA neurons only by an action on DA autoreceptors. Viceversa, in the unanesthetized animal LIS may influence DA neurons by simultaneously acting on DA autoreceptors and postsynaptic DA receptors.

The fact that in such preparation the effect of LIS is reminiscent of that of neuroleptics (Chiodo and Bunney, 1984; Mereu, 1982) suggests that the initial effect of the drug depends on an antagonistics interaction with postsynaptic DA receptors which triggers the compensatory activation of DA neurons via the striato-nigral feedback loop. This mechanism would precede and/or surpass the direct action of LIS on DA autoreceptors. The same mechanism might explain the reversal of APO-induced inhibition of neuronal firing, in the unanesthetized rats, and the ineffectiveness of LIS in the anesthetized preparation.

After a critical time period, LIS causes a long-lasting insensitivity of DA neurons to DA agonists and antagonists. This effect might be explained with a stable interaction of the drug with DA autoreceptors. According to this hypothesis, LIS might induce a stable conformational change of the receptors resulting in their partial activation. In turn, the occupancy of the receptors would prevent the effect of a full agonist such as APO, while the partial stimulation would prevent the excitatory response to neuroleptics.

Our results are in agreement with previous results obtained with bromocriptine, which was shown to inhibit DA synthesis and firing rate of DA neurons by interacting irreversibly with presynaptic and somato dendritic DA autoreceptors, respectively (Lieberman et al., 1983). We have reported that the potency of LIS to cause DA autoreceptor insensitivity remains unaltered after chronic treatment. However the sensitivity of such receptors returns to normal 24 after the last chronic treatment (Mereu et al., 1986 c).

The mixed agonist and antagonist properties of LIS might be relevant for its clinical applications. In Parkinsonian patients LIS is expected to act as agonist on "denervated" postsynaptic receptors (Clark et al., 1985). Indeed, chronic LIS has been proved to be an effective antiparkinsonian drug notable in having relatively fewer side effects with respect to other treatments (Lieberman et al., 1983; Gopinathan and Calne, 1981; Rinne, 1986; Seeman, 1981).

On the other hand in conditions associated with increased DA transmission, a partial agonist with high affinity for DA receptors might be expected to reduce DA transmission. In this regard, a compound like trans-dihydro-lisuride with weaker intrinsic activity than LIS (Krejci et al., 1985) might be a potentially effective antipsychotic.

Finally the long lasting insensitivity of DA autoreceptors after LIS might play a positive role in the clinical efficacy of a combination of LIS with L-dopa (Lieberman et al., 1983) in the treatment of parkinsonism. In fact, in these conditions, LIS should prevent l-dopa from inhibiting the activity of the spared DA neurons, via DA acting onto DA autoreceptors. Indeed the functional integrity of such neurons is considered to be important for preventing the on-off phenomena in parkinsonian patients (Gudrun and Paalzow, 1986; Rinne, 1986).

References

Bannon HJ, Grace AA, Bunney BS, Roth RH (1980) Evidence for an irreversible interaction of bromocryptine with central dopamine receptors. Naunyn Schmiedebergs Arch Pharmacol 312: 37–41

Chiodo LA, Bunney BS (1984) Effects of dopamine antagonists on midbrain dopamine cell activity. In: Usdin E, Carlsson A, Dahlstrom A, Engel J (eds) Catecholamines: neuropharmacology and central nervous system – theoretical aspects. A R Liss, New York, pp 369–391

Calne DB, Horowski R, McDonald RJ, Wuttke W (1983) Lisuride and other dopamine agonists. Raven Press, New York

Clark D, Hjorth S, Carlsson A (1985) Dopamine-receptor agonists: mechanisms underlying autoreceptor selectivity I. Review of the evidence. J Neural Transm 62: 1–52

Gessa GL, Westfall TC, Boi V, Mereu GP (1986 b) Mixed dopamine agonist and antagonist properties of lisuride. Federation Proceedings, 70th annual meeting. Abstract 1623, vol 45/3. p 433

Gessa GL, Yoon K-WP, Boi V, Westfall TC, Mereu G (1986 a) Different anesthesia preparations reveal appropriate responses of dopamine mid-brain neurons to dopamine agonist and antagonist drugs. Soc Neurosci Ann Meeting. Abstract 133.17, vol 12/1. p 486

Gopinathan G, Colne DB (1981) Actions of ergot derivatives in parkinsonism. In: Rose FC, Capildeo R (eds) Research progress in Parkinson's disease. Pitman Medical, New York, pp 324–332

Gudrun HM, Paalzow LK (1986) L-dopa: how it may exacerbate parkinsonian symptoms. Trends Pharmacol Sci 7: 15–19

Horowski R (1978) Differences in the dopaminergic effects of the ergot derivatives bromocriptine, lisuride, and d-LSD as compared with apomorphine. Eur J Pharmacol 51: 157–166

Kehr W (1984) Transdihydrolisuride, a partial dopamine receptor antagonist: effects on monoamine metabolism. Eur J Pharmacol 97: 111–119

Krejci I, Schuh J, Pragerova H, Dlabac A (1985) Lisuride and transdihydolisuride: differences in action on central dopaminergic functions in dependence on the location and the state of receptors. Pol J Pharmacol Pharm 37: 263–271

Lieberman AN, Goldstein M, Gopinathan G, Neophytides A, Leibwitz M, Walker R, Hiesiger E (1983) Lisuride in Parkinson's disease and related disorder. In: Calne DB, McDonald RJ, Horowski R, Wuttke W (eds) Lisuride and other dopamine agonists. Raven Press, New York, pp 419–430

Llinas R, Greenfield SA, Jahnsen H (1984) Electrophysiology of pars compacta cells in the in vitro substantia nigra: a possible mechanism for dendritic release. Brain Res 294: 127–132

Mereu GP (1982) Activation of dopaminergic neurons by sulpiride. In: Biggio G, Costa E, Gessa GL, Spano PF (eds) Receptors as supramolecular entities, vol 1. Pergamon Press, New York, pp 147–153

Mereu GP, Fanni B, Gessa GL (1984) General anesthetics prevent dopaminergic neuron stimulation by neuroleptics. In: Usdin E, Carlsson A, Dahlstrom A, Engel J (eds) Catecholamines: neuropharmacology and central nervous system-theoretical aspects. Neurobiology, vol 8 B. A R Liss, New York, pp 353–358

Mereu G, Muntoni F, Collu M, Boi V, Gessa GL (1986 a) Delayed blockade of dopamine autoreceptors by lisuride. In: Biggio G, Costa E (eds) Dopamine receptors physiology. Liviana Press, Padova Fidia Research Series, vol 3, pp 599–601

Mereu G, Muntoni F, Romani F, Calabresi P, Boi V, Gessa GL (1986 b) Responsiveness to "autoreceptor" doses of apomorphine is inversely correlated with the firing rate of dopaminergic neurons. Neurosci Lett 65: 161–166

Mereu G, Hu XT, Wang RY, Westfall TC, Gessa GL (1986 c) Failure of subchronic lisuride to modify A 10 dopamine autoreceptors sensitivity. Brain Res 408: 210–214

Rinne UK (1986) The importance of an early combination of a dopamine agonist and levodopa in the treatment of Parkinson's disease. In: van Maanen J, Rinne UK (eds) Lisuride: a new dopamine agonist and Parkinson's disease. Proceedings of a symposium held in Berlin. Excerpta Medica, Amsterdam, pp 64–71

Rogawski MA, Aghajanian GK (1979) Response of central monoaminergic neurons to lisuride: comparison with LSD. Life Sci 24: 1289–1298

Seeman P (1981) Brain dopamine receptors. Pharmacol Rev 32: 229–313

Skirboll LR, Grace AA, Bunney BS (1979) Dopamine auto- and postsynaptic receptors: electrophysiological evidence for differential sensitivity to dopamine agonists. Science 206: 80–82

Stoof JC, Kebabian JW (1984) Two dopamine receptors: biochemistry, physiology and pharmacology. Life Sci 35: 2281–2296

Tissari AH, Gessa GL (1983) Ergot-induced inhibition of dopamine synthesis in striatal synaptosomes: a D-2 DA receptor-mediated mechanism. In: Calne DB et al (eds) Lisuride and other dopamine agonists. Raven Press, New York, pp 33–34

Traub M, Wagner HR, Hassan M, Jackson-Lewis V, Fahn S (1985) The effects of chronic bromocriptine treatment on behaviour and dopamine receptor binding in the rat striatum. Eur J Pharmacol 118: 147–154

Walters JR, Baring MD, Lakoski JM (1979) Effects of ergolines on dopaminergic and serotoninergic single unit activity. In: Fuxe K, Calne DB (eds) Dopaminergic ergot derivatives and motor function. Pergamon Press, New York, pp 207–221

Walters JR, Baring MD, Carlsson JH (1986) Effects of D-1 and D-2 receptor stimulation on single unit activity in substantia nigra and globus pallidus. In: Woodruff GN (ed) Dopamine systems and their regulation. Macmillan Press, London, pp 63–78

White FJ, Wang RY (1983) Comparison of the effects of LSD and lisuride on A 10 dopamine neurons in the rat. Neuropharmacology 22: 669–676

Author's address: Dr. G. L. Gessa, Farmacologica Clinica, Universitaria de Cagliari, Via Porcell 4, I-09100 Cagliari, Italy.

J Neural Transm (1988) [Suppl] 27:211–218

Pharmacokinetic studies with sustained-release formulations of levodopa in healthy volunteers

M. Gerlach[1], B. Gebhardt[1], W. Kuhn[1], and H. Przuntek[2]

[1] Department of Neurology, University of Würzburg, Würzburg, and
[2] Department of Neurology, University of Bochum, St. Joseph-Hospital, Bochum,
Federal Republic of Germany

Summary. On the basis of the results of a large number of studies of the correlation between plasma levodopa levels and the occurrence of mobility disorders, it was concluded that side-effects such as dyskinesia, end-of-dose, peak-dose and on-off phenomena, which occur especially during levodopa long-term treatment of Parkinson patients, may be caused by fluctuations in the plasma levodopa levels. Some preliminary trials using sustained-release formulations were not up to expectations. Still, we studied the question of whether it might not be possible after all to obtain sustained plasma levodopa levels over prolonged periods of time with levodopa sustained-release (S. R.) formulations, without causing a high initial peak. For this purpose, we compared the pharmacokinetics of six different levodopa S. R. formulations with a standard preparation in two randomized cross-over trials in healthy male volunteers.

Although the levodopa S. R. formulations prolonged the time to peak blood levels in all cases, a plateau that was constant over a longer period of time could not be obtained. However, none of the S. R. preparations studied gave rise to a high initial peak of the plasma levodopa concentration. The levodopa concentrations had again returned to endogenous levels within no more than 8 hours of administration of all six levodopa S. R. formulations.

Introduction

Despite the occurrence of side-effects such as dyskinesias, on-off phenomena and psychoses, particularly during long-term treatment, in substitution therapy $(-)$-3-(3,4-dihydroxyphenyl)-L-alanine (levodopa) is still the principal drug in the management of Parkinson's disease. The molecular biochemistry causes of these side-effects have not yet been fully elucidated. Possible explanations include progressive degeneration of dopaminergic and other neurons (Hornykiewicz, 1977) with reduced vesicular ability to store dopamine (Leenders et al., 1986), and a possible supersensitization of the dopamine receptors and post-receptor mechanisms (Jenner et al., 1986).

The observation that the occurrence of mobility disorders frequently correlates with the time course of the plasma levodopa concentration led to the view that the side-effects such as dyskinesia, end-of-dose akinesia, peak-dose effect and on-off phenomenon, which occur particularly during long-term treatment, may be caused by large fluctuations in plasma levodopa levels (Fahn 1974; Sweet et al., 1974; Tolosa et al., 1975; Eriksson et al., 1984; Hardie et al., 1984; Nutt et al., 1984; Hardie et al., 1986). Although continuous intravenous infusion tends to reduce these movement impairments (Quinn et al., 1984; Hardie et al., 1986), chronic intravenous levodopa administration is practicable neither in the treatment of in-patients nor in out-patients. The introduction of levodopa sustained-release formulations that release levodopa slowly and continuously over an extended period of time, was therefore hoped to optimize Parkinson therapy (longer dosage interval, reduced incidence of dyskinesia and on-off phenomenon).

This study investigates the pharmacokinetics of five different levodopa S. R. formulations and one quick-release (Q. R.) formulation in comparison with one standard preparation in healthy volunteers in two randomized crossover trials. Comparison of the pharmacokinetic parameters may thus be helpful in determining which levodopa S. R. formulation is the best approach to the management of Parkinson patients.

Subjects and methods

Study design

This double-blind study was performed in two randomized cross-over trials in 6 healthy male volunteers in accordance with the recommendations of the declaration of Helsinki in its revised version of October 1975, the WHO guidelines (technical report series no. 563, 1976: Guidelines for the development of drugs for use in man), and the provisions of the German Drug Act of August 1976 (section 40).

The first trial compared the pharmacokinetics of three S. R. formulations (Batch no. 12/2–4) and one regular-release non-eneteric-coated levodopa formulation (12/1) with a standard preparation (12/5) up to 12 hours following administration of single 100 mg levodopa doses. A six-day wash-out period was observed before the next levodopa dose, so that the test preparations were all given on the same day of the week. The sequence of administration of the S. R. formulations, the Q. R. formulation and the standard preparation was randomized using the Latin square method.

The second trial was conducted one year later and used the same subjects who had participated in the first trial. In this study the pharmacokinetics of two novel S. R. formulations (11/7, 11/10) were determined in comparison with the 12/2 formulation that had already been tested in the first trial.

One day before each levodopa administration, the subjects were given 50 mg benserazide with 150 ml water at 8-hour intervals. On the day of the levodopa dose, the subjects were given 50 mg benserazide in the morning one hour before the levodopa dose and another 50 mg benserazide dose six hours after the first benserazide dose, each dose together with 150 ml water.

The subjects were fasted (except for soda water) at least 10 hours after levodopa administration.

Immediately before the levodopa dose and 10, 30, 45, 60, and 90 minutes as well as 2, 2.5, 3, 3.5, 4, 5, 6, 7, 8, 9, 10, and 12 hours after levodopa administration, 10 ml blood were sampled, prepared and analyzed as described in "Determination of levodopa levels".

Subjects

The 6 healthy male volunteers are between 26 and 34 years of age. Their weight ranges between 71 und 85 kg, their height between 1.66 und 1.90 m.

In terms of the sex-specific BROCA Index

$$BI = (\text{Weight in kg})/(\text{Height in cm} - 100) \times f$$
$$\text{where } f = 0.90 \text{ for men}$$

none of the subjects is obese (BI greater than 1.2).

Preparations

Capsules each containing 100 mg enteric-coated levodopa (Batch no. 12/2–4 and 11/7, 11/10, Degussa Pharma Gruppe, Chemiewerk Homburg) in five different sustained-release formulations.

Capsules containing 100 mg non-enteric-coated levodopa (Batch no. 12/1, Degussa Pharma Gruppe, Chemiewerk Homburg).

Capsules containing 100 mg quick-release levodopa as reference preparation (Batch no. 12/5, Degussa Pharma Gruppe, Chemiewerk Homburg).

Capsules containing 57 mg benserazide hydrochloride (N-(DL-seryl)-N'-(2,3,4-trihydroxybenzyl)hydrazine · HCL; Ro 04-4602/007, Batch no. GHK 080 CO 1 LO 124; Hoffmann-La Roche, Basel, Switzerland) as decarboxylase inhibitor.

Inclusion criteria

Subjects aged between 20 and 35 years not showing any abnormalities in their histories were admitted to the trials. Further criteria for inclusion into the study were: all clinical parameters inside the normal range; non-smoker or smoker of no more than 15 cigarettes per day; negative drug screening and negative alcohol test no earlier than seven days before the start of the trial; normal EEG, normal laboratory parameters no earlier than seven days before the trial; less than 20% deviation of the body weight from normal.

Exclusion criteria

Subjects in whom sympathomimetic drugs were contraindicated; individuals with acute diseases, laboratory parameters showing clinically relevant deviations from normal; subjects who had participated in a clinical trial during the past two months were excluded from the study. Subjects who had not given their informed consent to participate in the study or subjects who took other drugs during the study were also excluded.

Determination of levodopa levels

At fixed times, blood samples were obtained from the subjects in an horizontal position via an indwelling venous catheter. The blood was collected in EDTA-coated tubes containing 100 μl of a 0.5% sodium disulfite solution. The blood was centrifuged (2500 g for 10 min at 4 °C) and the plasma was divided into three aliquots and frozen immediately at −20 °C until analyses (all samples were analyzed within two weeks of collection). A 1000 μl volume of thawed sample was deproteinized by treatment with 150 μl of 0.01 M hydrochloric acid, 300 μl of 0.7 M perchloric acid and 10 μl of 0.1% EDTA solution. After centrifugation of this mixture at 100000 g and 4 °C, the supernatant was decanted. The further analytical procedures, described in an earlier report (Gerlach et al., 1986), can be summarized as follows: A 750 µl sample of the deproteinized plasma was loaded onto a C 18 cartridge (Sep-Pak C 18, Millipore, Eschborn, Federal Republic of Germany), which had previously been washed and equilibrated according to the manufacturer's instructions with 2 ml methanol followed by 5 ml clean water. Loading was carried out by means of a 2 ml disposable syringe and was completed within 2 min. The column was then washed with 1 ml water before eluting in less than 2 min with 1000 µl of 15% ethanol in water. A 20 µl aliquot of the ethanolic eluate was injected into the HPLC system. Chromatography was performed using a HPLC system from Milton-Roy (Hasselroth, Federal Republic of Germany) and Rheodyne injection valve with a 20 µl loop coupled to a Metrohm 656 (Filderstadt, Federal Republic of Germany) amperometric detector equipped with a glassy carbon electrode. A reverse-phase Nucleosil 120—5 C 18 column, (250 mm × 4.6 mm I. D.; Macherey-Nagel, Düren, Federal Republic of Germany), protected by a guard column containing Shandon Hypersil (5 µl, Bischoff Analysentechnik, Stuttgart, Federal Republic of Germany) and operated at room temperature was used. The elution buffer of pH 3.35 contained 84% 0.1 M sodium dihydrogen phosphate and 16% methanol with 2.6 mM sodium octylsulfonate, 0.1 mM disodium EDTA and 0.25 mM triethylamine. Isocratic elution was carried out at a flow-rate of 1.5 ml/min. The detector potential was set at 0.75 V versus the reference electrode Ag/AgCl at a sensitivity of 10—50 nA/V.

Test criteria and statistical analysis

As a result of the non-monotonous course of the plasma levodopa level curves after reaching the peak, which has to be interpreted as biological scatter, statistical analysis was performed using the median rather than the mean. The following pharmacokinetic parameters were determined: Maximum plasma levodopa concentration (C_{max}) and the time to peak level (t_{max}). The total area under the plasma concentration-time curve (AUC) was estimated by using the trapezoidal rule. Given that levodopa is a naturally occurring compound, the pre-dose level (Co) was used to correct the AUC. The relative bioavailability of levodopa, i.e., the ratio of the AUC values of the new formulations and the standard preparation was determined by measuring the AUC values.

As a criterion for assessing the extent of the prolongation of the effect of the S. R. formulations, the levodopa half-value duration was also calculated. The halfvalue duration of plasma level curves that rise in a monotonous fashion until reaching C_{max} to fall again in a monotonous manner is defined as the time during which the plasma level is greater than or equal to $C_{max}/2$ line (Meier et al., 1981). In calculating the plateau curve, only those subjects were considered whose concentration curve showed exactly two intersections with the $C_{max}/2$ line.

Since the two pharmacokinetic trials were conducted in the same subjects and since there was good agreement between the two most important parameters, C_{max} and t_{max}, the results of the six identical subjects are compared below.

Results

Figure 1 shows the medians of the plasma levodopa concentrations following administration of the 5 S. R. formulations, the standard preparation and the Q. R. formulation 12/1. It becomes apparent that the S. R. formulations prolong the time to peak levodopa level in blood. On the S. R. formulations 12/2–4, 11/7, and 11/10, there are no well-defined peaks as are seen with the standard preparation and the formulation 12/1. On the other hand, a continuous rise in

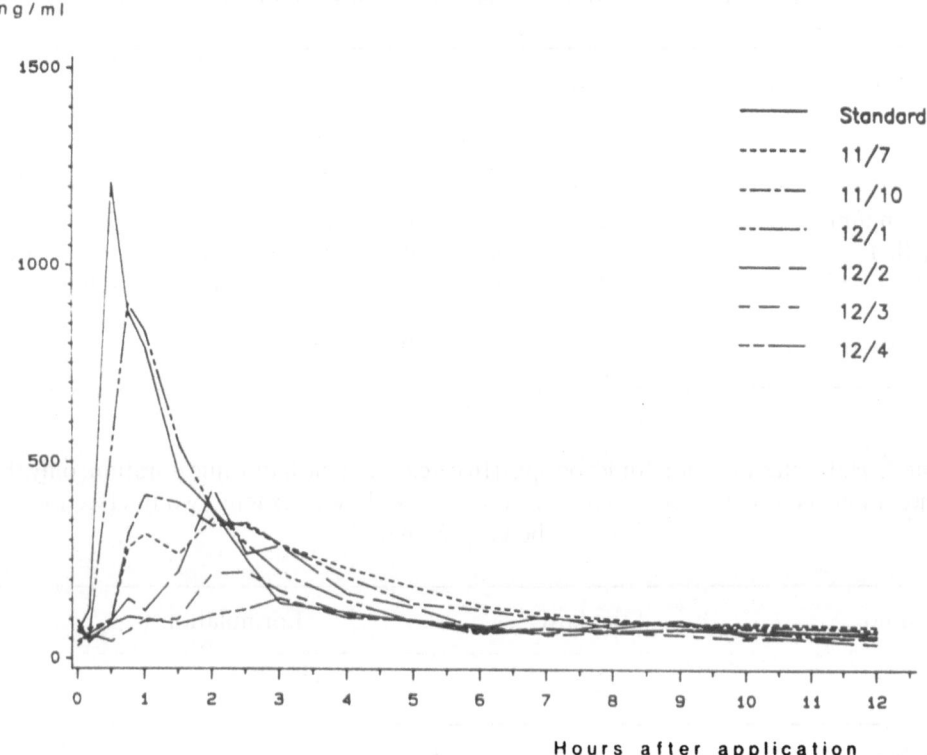

Fig. 1. Plasma levodopa concentrations (medians) over time following oral administration of various S. R. formulations and a standard preparation (in each case 100 mg levodopa + decarboxylase inhibitor)

levodopa concentration is not obvious but only roughly discernible for formulations 11/7, 11/10, and 12/2 for varying durations. For all preparations studied, the endogenous levodopa baseline level is reached again after 8 hours.

The pharmacokinetic parameters of levodopa in the formulations studied are summarized in Table 1. This shows that the Q. R. formulation 12/1 differs only slightly from the standard in any parameter. In contrast, the formulations 12/2–4 show clearly lower C_{max} values, with the corresponding t_{max} values being definitely increased. The C_{max} values for the formulations 11/7 and 11/10 are about the same as the level for 12/2. However, 11/7 ist clearly distinguished from 12/2 by its biphasic plasma levodopa concentration. The relative bioavailability of formulations 12/2–4 falls from 39% to 2%. The bioavailabilities of formulations 11/7 and 11/10 are 71% and 82%, respectively.

For assessing the S. R. formulations, the half value durations were also calculated. Table 2 summarizes the half value durations for the formulations 12/1, 12/2, 11/7, 11/10 and the standard. As a result of its flat levodopa peak, half value duration was not calculated for formulation 12/3 and 12/4. Comparison of the available data with the corresponding C_{max} values shows a clear relationship between half value duration and C_{max} for formulations 12/1 and 12/2: The smaller the C_{max}, the longer the half value duration. The half value

Table 1. Pharmacokinetic parameters of levodopa (medians)

Batch no.	Standard	Formulation					
	12/5	12/1	12/2	12/3	12/4	11/7	11/10
C_{max} (ng/ml)	1,242	952	490	297	162	494	526
t_{max} (hr)	0.50	0.88	2.00	2.50	4.50	2.00	1.00
AUC (ng × hr/ml)	1,945	1,703	575	755	44	1,136	1,322
Rel. bioavail-ability (in %)	100	88	30	39	2	71	82

Table 2. Half value duration for levodopa. (In calculating the half value duration, only those subjects were considered whose concentration curve showed exactly two intersections with the $C_{max}/2$ line)

Batch no.	Standard	Formulation			
	12/5	12/1	12/2	11/7	11/10
Number	6	5	5	5	6
Median (min)	48.0	71.5	101.8	164.8	196.4

durations of formulations 11/7 and 11/10 (164.8 min and 196.4 min) are clearly longer than that of formulation 12/2.

Discussion

The pharmacokinetic studies of the various levodopa formulations show that S. R. formulations fail to maintain constant plasma levodopa levels over prolonged periods of time. After as little as 6–8 hours of administration of the preparations studied, the levodopa concentrations have returned to almost endogenous levels. These findings are supported by the pharmacokinetic studies of other sustained-release levodopa preparations (Marion et al., 1986; Mena et al., 1986; Nutt et al., 1986; Juncos et al., 1987).

Nor can plasma levodopa levels be kept constant over prolonged periods of time by continuous intravenous infusion of levodopa, accompanied by oral administration of a decarboxylase inhibitor, since peripheral levodopa is metabolized predominantly by catechol-O-methyl-transferase (COMT) to 3-O methyldopa (3-OMD) and rapidly by transamination to Vanilpyruvate, Vanillacetate, and 2,4,5-trihydroxyphenylacetic acid (Nutt and Fellmann, 1984).

In addition to fast enzymatic metabolism, absorption from the gastrointestinal tract is also an important factor in oral levodopa Parkinson therapy.

Like other aromatic amino acids, levodopa, once released from the stomach, is absorbed from the upper part of the duodenum by an active transport mechanism that is subject to saturation (Gundert-Remy et al., 1983); what levodopa is not absorbed is additionally oxidatively degraded to dopaquinone (Nutt and Fellmann, 1984; Gerlach et al., 1986).

It is therefore appreciable that, in the case of S. R. formulations, the duration of levodopa release and the time to peak levodopa levels are limited by the (relatively short) time the preparation remains in the gastrointestinal tract. In fact, the limited time levodopa remains in the upper gastrointestinal tract and the quick metabolism are responsible for the finding that the S. R. formulations 12/2–4, 11/7, and 11/10 show a lower relative bioavailability than the quick-release standard preparation.

On the basis of the available pharmacokinetic parameters, formulation 11/10 would appear to be the most suitable candidate for a trial in Parkinson therapy: In comparison to the other S. R. formulations, it peaks after a short time yet without reaching extremely high initial concentrations, and it shows a considerable longer half value duration as well as the smallest decrease in relative bioavailability. These properties suggest that this formulation may optimize Parkinson therapy (reduction in the number of daily doses, longer intervals between doses, reduction in the side-effects described above).

Acknowledgements

We would like to thank Dr. H. Nowak (Department of Biometry, Degussa Pharma, Frankfurt, Federal Republic of Germany) for calculating the pharmacokinetic data.

References

Eriksson T, Magnusson T, Carlsson A, Granérus AK (1984) "On-off" phenomenon in Parkinson's disease: correlation to the concentration of Dopa in plasma. J Neural Transm 59: 229–240

Fahn S (1974) "On-off" phenomenon with levodopa therapy in parkinsonism. Neurology 24: 431–441

Gerlach M, Klaunzer N, Przuntek H (1986) Determination of L-Dopa and 3-O-methyl-Dopa in human plasma by extraction using C_{18} cartridges followed by high-performance liquid chromatographic analysis with electrochemical detection. J Chromatogr 380: 379–385

Gerlach M, Gebhardt B, Kuhn W, Przuntek H (1986) Die Abhängigkeit der Resorption des L-Dopa von Galenik und veränderter Magensaftsekretion des Parkinson-Patienten. In: Fischer PA (ed) Spätsyndrome der Parkinson-Krankheit. Editiones Roche, Basel, pp 271–279

Gundert-Remy UR, Hildebrandt R, Stiehl A, Weber E, Zürcher G, Da Prada M (1983) Intestinal absorption of levodopa in man. Eur J Clin Pharmacol 25: 69–72

Hardie RJ, Lees AJ, Stern GM (1984) On-off fluctuations in Parkinson's disease. Brain 107: 487–506

Hardie RJ, Malcolm SL, Lees AJ, Stern GM, Allen JG (1986) The pharmacokinetics of intravenous and oral levodopa in patients with Parkinson's disease who exhibit on-off fluctuations. Br J Clin Pharmacol 22: 429–436

Hornykiewicz O (1977) Biogenic amines in the central nervous system. In: Vinken PJ, Bruyn GW (eds) Handbook of clinical neurology vol 29, part 3. North-Holland, Amsterdam, pp 495–483

Jenner P, Boyce S, Marsden CD (1986) Effect of repeated L-Dopa administration on striatal dopamine receptor function in the rat. In: Fahn S et al (eds) Recent developments in Parkinson's disease. Raven Press, New York, pp 189–203

Juncos JL, Fabborini G, Mouradian MM, Serrati C, Kask AM, Chase TN (1987) Controlled release levodopa treatment of motor fluctuations in Parkinson's disease. Neurol Neurosurg Psychiatry 50: 194–198

Leenders K, Palmer A, Turton D, Quinn N, Firnau G, Garnett S, Nahmias C, Jones T, Marsden CD (1986) Dopa uptake and dopamine receptor binding visualized in the human brain in vivo. In: Fahn S et al (eds) Recent development in Parkinson's disease. Raven Press, New York, pp 103–113

Marion MH, Stocchi F, Quinn NP, Jenner P, Marsden CD (1986) Single-dose study of slow release preparation of levodopa and benseracide (Madopar HBS) in Parkinson's disease. In: Yahr MD, Bergmann KJ (eds) Advances in neurology, vol 45. Raven Press, New York, pp 493–496

Meier J, Rettig H, Hess H (1981) Biopharmazie-Theorie und Praxis der Pharmakokinetik. G Thieme, Stuttgart New York

Mena MA, Muradas V, Bazan E, Reiviz J, de Yebenes JG (1986) Pharamcokinetics of L-Dopa in patients with Parkinson's disease. In: Yahr MD, Bergmann KJ (eds) Advances in neurology, vol 45. Raven Press, New York, pp 481–486

Nutt JG, Fellman JH (1984) Pharmacokinetics of Levodopa. Clin Neuropharmacol 7: 35–49

Nutt JG, Woodward WR, Hammerstad JP, Carter JH, Anderson JL (1984) The "On-off" phenomenon in Parkinson's disease. N Engl J Med 310: 483–488

Nutt JG, Woodward WR, Carter JH (1986) Clinical and biochemical studies with controlled-release levodopa/carbidopa. Neurology 36: 1206–1211

Quinn N, Parkes JD, Marsden CD (1984) Control of on/off phenomenon by continuous intravenous infusion of levodopa. Neurology 34: 1131–1136

Sweet RD, Fletcher H, Dowell MC (1974) Plasma dopa concentrations and the "on-off" effect after chronic treatment of Parkinson's disease. Neurology 24: 953–956

Tolosa ES, Martin WE, Cohen HP, Jacobsen RL (1975) Patterns of clinical response and plasma dopa levels in Parkinson's disease. Neurology 25: 177–183

Authors' address: Prof. Dr. H. Przuntek, Department of Neurology, University of Bochum, St. Joseph-Hospital, Gudrunstrasse 56, D-4630 Bochum, Federal Republic of Germany.

J Neural Transm (1988) [Suppl] 27: 219–225

Parkinson's disease and PET tracer studies

K. L. Leenders

MRC Cyclotron Unit, Hammersmith Hospital, London, U.K.

Summary. Although the cause of Parkinson's disease is still unknown it has become clear that dopaminergic cell loss in the substantia nigra is a prominent feature of the disease. The study of cerebral dopaminergic function – dominated by the nigrostriatal system – in relation to motor abnormalities and their response to treatment has been in the centre of research for many years. However, until recently it was not possible to approach metabolism or neurotransmitter function in brain directly in vivo. To be able to do so seems highly necessary since Parkinson's disease is a slowly progressive condition and one wishes to know how cerebral functions are deranged in the beginning of the disease.

Positron emission tomography (PET) using specific tracers (labelled with short-lived radionuclides like oxygen-15, carbon-11 or fluorine-18) can measure certain physiological or biochemical functions focally in brain directly in patients or healthy volunteers (for an extensive overview, see Phelps, 1986).

Energy metabolism and cerebral blood flow (CBF)

Sequential inhalation of trace doses of oxygen-15 labelled CO_2 and O_2 allows measurement of regional CBF, oxygen extraction and oxygen utilization. These measurements give an impression of the overall neuronal functional level in a region of the brain. Focal and global abnormalities in different stages of the disease have been detected (Wolfson et al., 1985) and also responses to levodopa therapy have been studied (Leenders et al., 1985 a).

Striatal dopamine functions

Since the nerve terminals from the dopaminergic nigro-striatal pathway are highly concentrated in striatum, native dopamine and related enzymes are equally abundant in that region. This circumscribed focal abundance of dopaminergic function makes PET tracer measurements easier. Several tracers, each specific for a particular aspect of the dopaminergic system, have been developed in recent years and are being validated further.

L-[^{18}F]fluorodopa

Cerebral L-dopa uptake in man can be demonstrated with PET using fluorine-18 labelled L-fluorodopa (F-dopa) (Fig. 1) as shown for the first time by Garnett

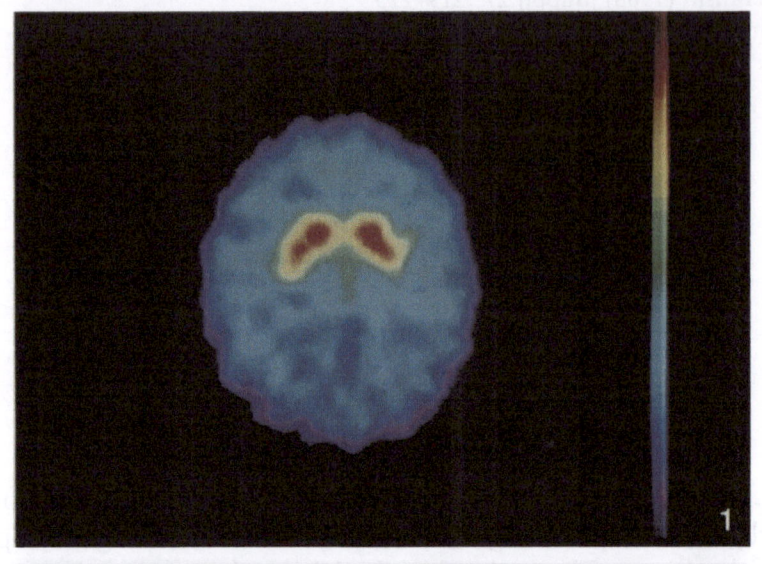

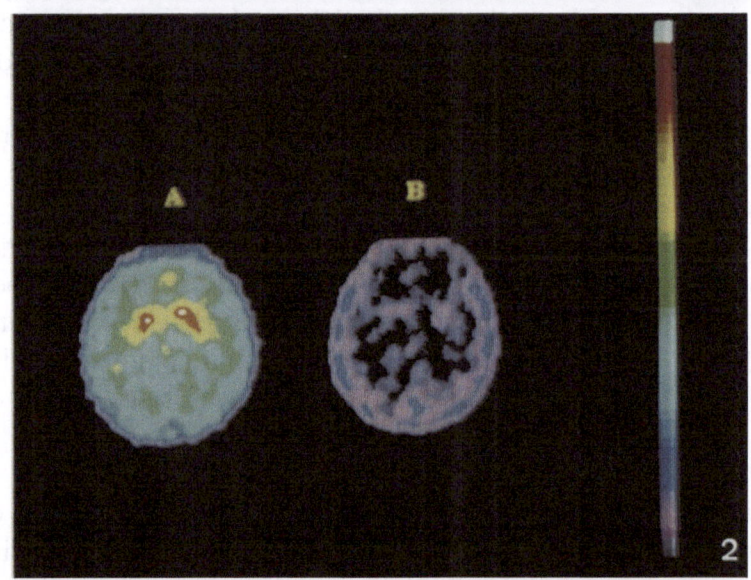

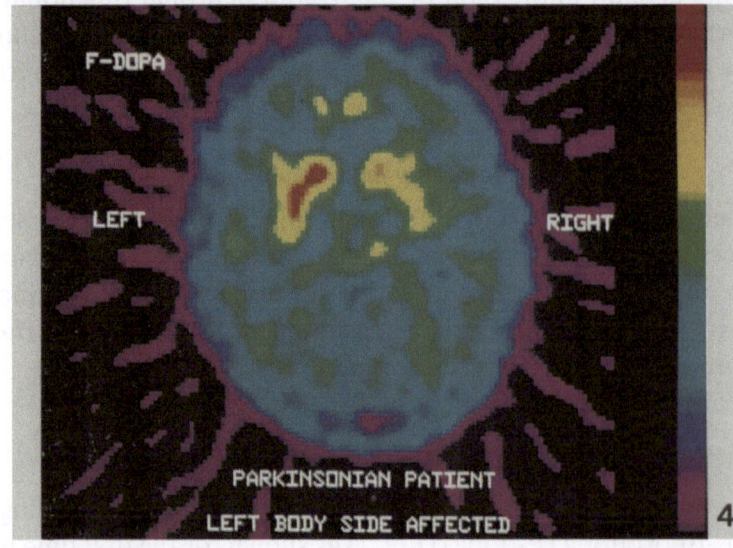

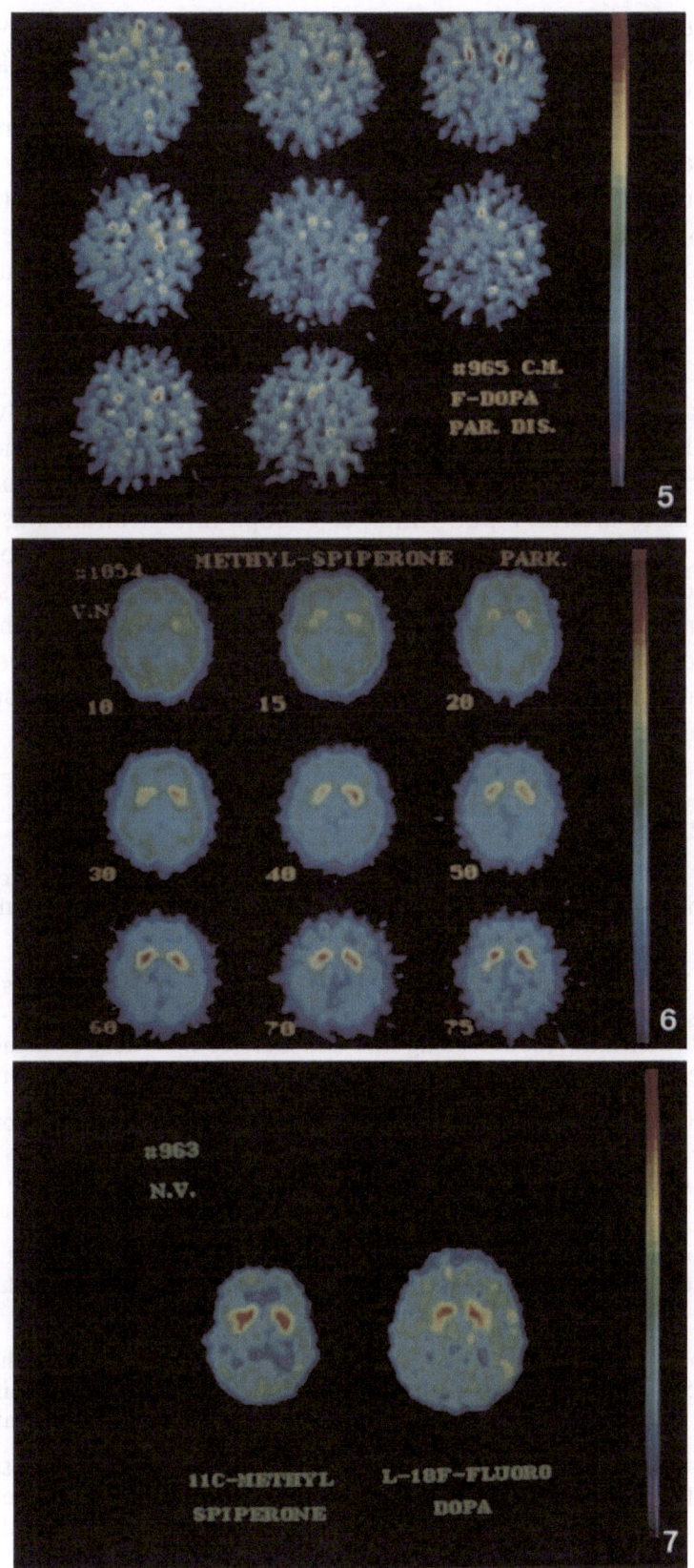

and colleagues. After intravenous injection of a bolus the radioactivity is at first diffusely distributed in the body and brain, but gradually (Figs. 2 and 3) a specific uptake pattern becomes obvious: the main retention of tracer is seen in striatal regions of the brain, whereas in surrounding non-dopaminergic brain tissue a gradual but steady washout of radio-activity takes place. Since PET only can measure the total number of radionuclide disintegrations (in this case fluorine-18) in a volume of tissue, assumptions need to be made about the chemical form of the measured radiotracer. It seems reasonable to assume that in the case of F-dopa the activity in striatum mainly represents fluorinated dopamine, at least in the first part of the studies (Garnett et al., 1983; Leenders et al., 1986 a).

In Parkinson's disease uptake of F-dopa in striatum is markedly decreased as expected (Nahmias et al., 1985; Leenders et al., 1986 a) (Figs. 4 and 5). Influx of tracer can be assessed by calculating an influx constant comparing local tissue uptake to arterial input. Influx of F-dopa into striatum was negatively correlated with severity of bradykinesia in patients with Parkinson's disease (Leenders et al., 1986 b). On the other hand simple ratios between striatal and non-dopaminergic brain regions can distinguish F-dopa uptake in healthy volunteers and patients in different stages of the disease (Leenders et al., 1986 a). Equally, this may be sufficient to detect possible exposure to the dopaminergic neurotoxin MPTP in man (Calne et al., 1985).

Fig. 1. F-dopa uptake in a cross-section through both striata of the brain of a normal volunteer. The image is a summation of all the scans (after correction for attenuation) between 100 and 220 min after administration of the tracer. In this and the following figures the top of the images indicate the front of the brain, viewed from above (with permission from J Neurol Neurosurg Psychiatry 49: 853–860)

Fig. 2. Two images depicting accumulated radio-activity (60 to 100 minutes) in a cross-section of the brain of a healthy volunteer after F-dopa administration. **A** Base-line (fasting) condition. **B** After intravenous loading with a mixture of eight neutral amino acids. The colour scale is identical for both images (with permission from Trends Neurosci 9: 1–5)

Fig. 4. F-dopa uptake in the brain of an untreated moderately affected parkinsonian patient, whose left body side was predominantly affected. The right striatum shows less F-dopa uptake then the left, although both sides have diminished values compared to control level (data not shown here)

Fig. 5. F-dopa uptake in the brain of a severely affected parkinsonian patient. The images are 10 minute consecutive measurements between 1 and 2 hours after tracer administration. Hardly any striatal F-dopa uptake is visible

Fig. 6. ^{11}C-methyl-spiperone uptake in the brain of a parkinsonian patient. The images are consecutive 5 and 10 minute measurements from 10 to 75 minutes after administration of the tracer. Rapid specific accumulation of radio-activity is seen in the striatal regions

Fig. 7. Brain uptake of the tracers ^{11}C-methyl-spiperone and F-dopa in the same healthy volunteer. The image visualizes radio-activity distribution in a horizontal cross-section at the level of the striatum. Uptake of both tracers is highest in striatum (with permission from Lancet 1984: 110–111)

L-[¹⁸F] fluorodopa uptake

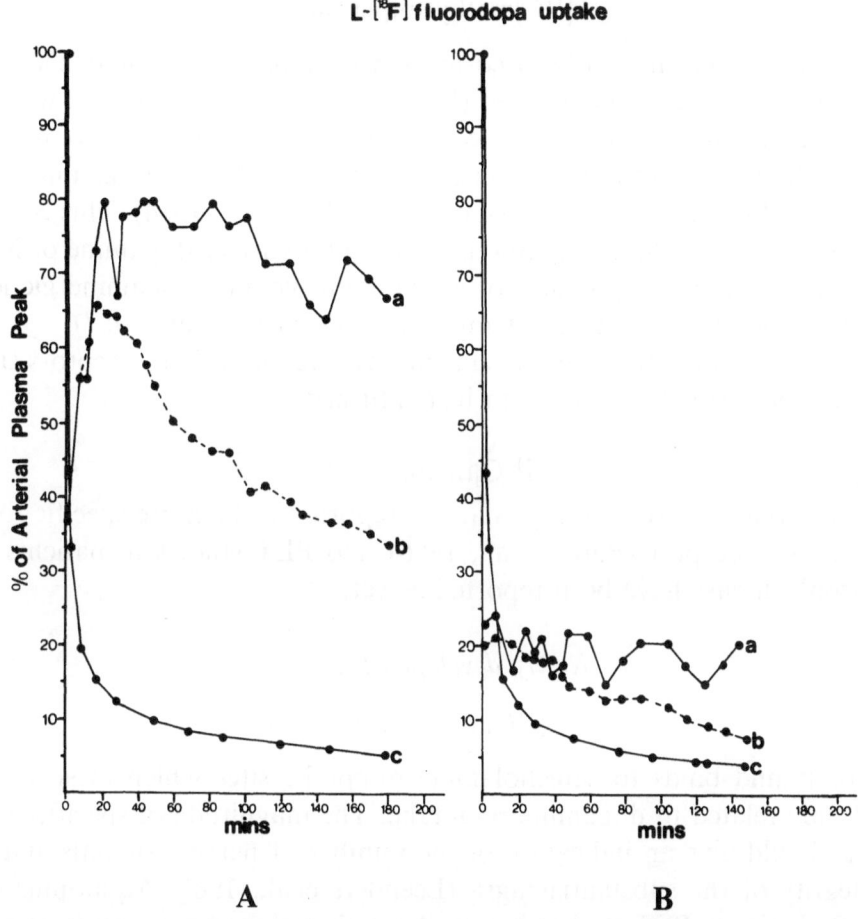

Fig. 3. Graphical display of the measured radio-activity as visualized in Fig. 2. The tissue values are expressed as a percentage of arterial plasma peak and then for clarity multiplied with a factor 10. *a* Striatum; *b* non-dopaminergic surrounding brain; *c* plasma concentration of radionuclide. Plasma radio-activity is similar before (**A**) and after (**B**) amino-acid loading. However, tissue uptake is markedly reduced in all regions (with permission from Trends Neurosci 9: 1–5)

F-dopa can also be used to demonstrate in vivo competition at the blood-brain barrier with large neutral amino acids (Leenders et al., 1986c). This is illustrated in Figs. 1 and 2.

Dopamine receptor ligands

Several compounds belonging to the neuroleptics (predominantly dopamine D_2 antagonists) have been used to assess dopamine receptor binding, particularly in striatum where these receptors are in highest concentration. The main aim is to try to deduce from this the number of receptors in a certain volume of tissue. There are many theoretical and practical considerations concerning this issue which cannot be discussed here (Kilbourn and Zalutski, 1985).

[^{11}C]methylspiperone

This tracer was the first to be used in man (Wagner et al., 1983). Decreased binding with age was demonstrated (Wong et al., 1984). In patients with Parkinson's disease no obvious differences in specific striatal binding compared to healthy volunteers were found (Leenders et al., 1985 b) (Fig. 6) as might have been expected from post-mortem studies (Bokobza et al., 1984). This contrasts with acute lesions of the substantia nigra using 6-hydroxy-dopamine or MPTP in animals where in the presence of a severe presynaptic dopaminergic lesion D$_2$ receptor binding is increased temporarily (Leenders et al., 1987).

Possibly levodopa treatment results in down regulation of receptors (Leenders et al., 1985 b) but this needs to be confirmed.

[^{11}C]raclopride

This tracer was recently developed and is reported to be more specific for the dopamine D$_2$ receptor (Farde et al., 1986). No PET studies in patients with Parkinson's disease have been reported as yet.

Newly developed tracers

[^{11}C]nomifensine

This compound binds to catecholamine re-uptake sites which in striatum is almost only related to dopamine re-uptake. The magnitude of specific striatal binding should give an indication of the number of nerve terminals and thus the integrity of the substantia nigra (Leenders et al., 1987; Aquilonius et al., 1987). Preliminary PET studies in a unilaterally lesioned monkey (using intra carotid MPTP) showed a 80 to 90% decrease of specific binding in the striatum on the lesioned side (Leenders et al., 1987). The first human PET studies comparing healthy volunteers and patients with Parkinson's disease seem to confirm these findings.

[^{11}C]deprenyl

This compound binds irreversibly to monoamine oxidase (type B) enzyme which is the main catabolic enzyme for dopamine in striatum. The active (L) and inactive (D) enantiomer of deprenyl can be labelled separately. Sequential administration and measurement of brain uptake by PET of both tracers allow to measure regional specific binding to the enzyme pool (Fowler et al., 1987). Preliminary PET studies in man show no obvious differences between healthy volunteers and patients with Parkinson's disease.

Concluding remarks

The complexity of PET tracer techniques allows only slow progress in tracer validation and subsequent application. However, it is expected that in vivo measurement of functional disturbances of the basal ganglia will contribute to the understanding of pathophysiology of movement disorders.

References

Aquilonius S, Bergstrom K, Eckernas S et al (1987) In vivo evaluation of striatal dopamine re-uptake sites using C-Nomifensine and positron emission tomography. Acta Neurol Scand 76: 283–287

Bokobza B, Ruberg M, Scatton B et al (1984) (^{3}H)spiperone binding, dopamine and HVA concentrations in Parkinson's disease and supranuclear palsy. Eur J Pharmacol 99: 167–175

Calne DB, Langston JW, Martin WRW et al (1985) Positron emission tomography after MPTP: observations relating to the cause of Parkinson's disease. Nature 317: 246–248

Farde L, Hall H, Ehrin E, Sedval G (1986) Quantitative analysis of dopamine-D-2 receptor binding in the living human brain using positron emission tomography. Science 231: 258–261

Fowler J, MacGregor R, Wolf A et al (1987) Regional distribution of monoamine oxidase A and B in human brain using ^{11}C-suicide inhibitors and positron emission tomography. Science 235: 481–485

Garnett ES, Firnau G, Nahmias C (1983) Dopamine visualized in the basal ganglia of living man. Nature 305: 137–138

Kilbourn MR, Zalutski MR (1985) Research and clinical potential of receptor based radiopharmaceuticals. J Nucl Med 26: 655–662

Leenders KL, Wolfson L, Gibbs JM, Wise RJS, Causon R, Jones T, Legg NJ (1985 a) The effects of L-dopa on regional cerebral blood flow and oxygen metabolism in patients with Parkinson's disease. Brain 108: 171–191

Leenders KL, Herold S, Palmer A et al (1985 b) Human cerebral dopamine system measured in vivo using PET. J Cereb Blood Flow Metab 5 [Suppl]: 517–518

Leenders KL, Palmer AJ, Quinn N et al (1986 a) Brain dopamine metabolism in patients with Parkinson's disease measured with positron emission tomography. J Neurol Neurosurg Psychiatry 49: 853–856

Leenders KL, Palmer A, Turton D et al (1986 b) Recent developments in Parkinson's disease. Raven Press, New York, pp 103–113

Leenders KL, Poewe WH, Palmer A et al (1986 c) Inhibition of L-dopa uptake into the human brain by amino acids demonstrated by positron emission tomography. Ann Neurol 20: 258–262

Leenders KL, Aquilonius S, Bergstrom et al (1987) Unilateral MPTP lesion in a rhesus monkey: effics on the striatal dopaminergic system measured in vivo with PET using various novel tracers. Brain Res (in press)

Nahmias C, Garnett ES, Firnau G, Lang A (1985) Striatal dopamine distribution in Parkinsonian patients during life. J Neurol Sci 69: 223–230

Phelps ME, Mazziotta JC, Schelbert HR (eds) Positron emission tomography and autoradiography. Principles and applications for the brain and heart. Raven Press, New York

Wagner HN, Burns HD, Dannals RF et al (1983) Imaging dopamine receptors in the human brain by positron tomography. Science 221: 1264–1266

Wolfson L, Leenders KL, Brown L, Jones T (1985) Alterations of regional cerebral blood flow and oxygen metabolism in Parkinson's disease. Neurology 35: 1399–1405

Wong DF, Wagner HN, Dannals RF et al (1984) Effects of age on dopamine and serotonin receptors measured by positron tomography in the living human brain. Science 226: 1393–1396

Author's address: Dr. K. L. Leenders, MRC Cyclotron Unit, Hammersmith Hospital, Du Cane Road, London W12, U.K.

J Neural Transm (1988) [Suppl] 27: 227–235

Mapping dopamine receptors in the human brain

J. M. Palacios[1], **M. Camps**[1]*, **R. Cortés**[1], and **A. Probst**[2]

[1] Preclinical Research, Sandoz Ltd, and [2] Department of Neuropathology, Institute of Pathology, University of Basle, Basle, Switzerland

Summary. We have investigated the anatomic localization of dopamine D_1 and D_2 receptors in the human brain using selective high affinity ligands for both types of dopamine receptors and the technique of receptor autoradiography. Dopamine D_1 receptors were labeled in postmortem human brain tissue sections using the antagonist [^{3}H]SCH 23390. Dopamine D_2 receptors were labeled in consecutive tissue sections using the agonist [^{3}H]205-502 and the antagonist [^{3}H]spiroperidol. D_1 and D_2 dopamine receptors presented a heterogeneous distribution in the human brain. The highest concentrations of both D_1 and D_2 receptors were found in parts of the basal ganglia, particularly the nucleus caudatus and putamen. Lower concentrations were seen in other areas for example, the lateral globus pallidus was enriched in D_2 receptors and the medial globus pallidus in D_1 receptors. The substantia nigra contained intermediate densities of both D_1 and D_2, D_1 receptors being present in higher concentrations. Dopamine D_1 receptors were also localized in areas outside of the basal ganglia, particularly in the neocortex, amygdala and hippocampal formation. Dopamine D_2 receptors were also present in areas outside of the basal ganglia, the most significant densities being found in the hippocampal formation. We observed a marked age-dependent decline in the density of D_1 receptors during the first decades of life. In contrast, D_2 receptor concentrations appeared to be unaltered with age. The distribution and densities of dopamine receptors were examined in 12 cases of Parkinson's disease and compared to a control adult population. No significant differences in density and distribution were seen for either D_1 nor D_2 receptors.

Introduction

The discovery by Hornykiewicz and his colleagues in the earlier '60 of a dramatic loss of dopamine in the nucleus caudatus of Parkinson patients represented the starting point for a large research effort to understand the role of this amine

---

* Visiting student. Supported by PGT grant of the European Science Foundation. Permanent Address: Department of Biochemistry, School of Medicine, University of Barcelona, E-08034 Barcelona, Spain.

in the human brain (see Hornykiewicz, 1972). One of the milestones in the research was the discovery of dopamine receptors, the site of action for important therapeutic agents such as neuroleptics and antiparkinsonian agents (Carlsson and Lindquist, 1963).

Two types of dopamine receptors are present in the central nervous system and in the periphery, the so-called D_1 and D_2 receptor subtypes according to Kebabian and Calne (1979). These two types of receptors were first differentiated on the basis of biochemical assays, particularly the dopamine stimulated adenylate cyclase and radioligand bindings assays (Spano et al., 1977), and later by the different physiological actions which they mediate (Stoof and Kebabian, 1984).

The postulation of the two subtypes of dopamine receptors has been essential for the development of specific pharmacological agents able to mimic or block the action of dopamine at these two types of receptors. Selective dopamine D_1 agents are the agonist SKF 38393 and the antagonist SCH 23390. Selective D_2 agents are quinpirole (LY 171555) and some neuroleptics, particularly those of the benzamide class. The question of where dopamine receptors are localized in the brain naturally followed the discovery of these receptors. It was only after the development of autoradiographic techniques that this question could be addressed. Both D_1 and D_2 dopamine receptors have been visualized in the laboratory animals using receptor autoradiography after in vivo and in vitro labeling of the tissues (Palacios and Pazos, 1986). Furthermore, ligands for dopamine receptors which can be used for the in vivo visualization in the living human being using non-invasive brain imaging techniques such as positron emission tomography have been developed. In this way dopamine D_2 receptors have been studied in the living human brain (Wagner et al., 1983; Baron et al., 1984; Wong et al., 1984; Friedmann et al., 1985; Farde et al., 1986).

The possible alterations of dopamine receptors in the pathological human being have also been object of an intensive research in the past ten years (Lee et al., 1978; Owen et al., 1978; Crow et al., 1978; Rinne et al., 1980; Bird and Iversen, 1982; Ruberg et al., 1985; Baron et al., 1985; Raisman et al., 1985; Pimoule et al., 1985). One of the main drawbacks of the methods used until now to examine the distribution of dopamine receptors in the human brain is the low anatomical resolution of the techniques employed. This is the reason why we have used light microscopic autoradiographic techniques to examine the distribution of both D_1 and D_2 receptors in the normal human brain (Camps et al., 1987; Cortés et al., 1987) and in cases suffering from Parkinson's disease.

In the present work we review our recent results on the localization of dopamine receptors in the human brain.

Methods

The general procedure used in our laboratory for mapping of receptors in human brain postmortem material has been recently reviewed (Palacios et al., 1986).

Data presented in this study are from 24 control cases, 14 males and 10 females, aged from 6 to 84 years old and with a mean postmortem delay of 9.17 ± 7.5 and 12 parkinsonian cases, 7 males and 5 females, aged from 71 to 95 years old and with a mean postmortem delay of 14.8 ± 9. We have also examined one case of progressive supranuclear palsy, a male aged 82 and with a postmortem delay of 31 h.

Dopamine receptor labeling for autoradiography

Labeling of D_1 receptors was carried out by incubating slide-mounted 10 μm thick tissue sections with 1 nM [³H]SCH 23390 in 50 mM tris-HCl, 120 mM NaCl, 5 mM KCl, 2 mM $CaCl_2$, 1 mM $MgCl_2$ (pH 7.4) for 45 min at room temperature (Dawson et al., 1986). Adjacent sections were incubated in the same medium in the presence of $10 - ^6$ M cis-flupentixol to determine the non-specific binding. After incubation sections were washed for 5 min in fresh cold buffer and quickly dried under a stream of cold air.

Labeling of D_2 receptors was carried out by using the agonist [³H]205-502 and the tritiated antagonist [³H]spiroperidol. Sections were preincubated in 170 mM Tris-HCl buffer (pH 7.5) for 30 min at room temperature and then incubated for 90 min with 1 nM [³H]205-502 in the same buffer at room temperature. After two washes in ice-cold buffer of 1 min each and dipping in ice-cold distilled water, slides were rapidly dried under a stream of cold air. Blanks were obtained using 1 μM of (+)butaclamol.

For [³H]spiroperidol autoradiography, slide mounted tissue sections were incubated at room temperature for 30 min with 0.4 nM [³H]spiroperidol in 170 mM tris-HCl buffer (pH 7.5) containing 120 mM NaCl, 5 mM KCl, 2 mM $CaCl_2$, 1 mM MgCl. After incubation slides were washed for 10 min (2 washes of 5 min each) at 4 °C and dipped in distilled water before drying. To block [³H]spiroperidol binding sites other than D_2 bindings sites, 1 μM ketanserin or haloperidol were added to the incubation medium. Blanks for D_2 binding sites were obtained using the dopamine D_2 antagonist sulpiride (1 μM).

Autoradiograms were generated by apposing the labeled tissues to a tritium-sensitive film ([³H]Ultrofilm, LKB) as described by Unnerstall et al. (1982). Films were analyzed using a computerized image-analysis system and appropriate radioactive standards as we have previously described (Cortés et al., 1984).

Results and discussion

The densities of dopamine D_1 and D_2 receptors we found in some areas of the neurologically normal human brain are summarized in Table 1. Higher densities of D_1 receptors compared to D_2 receptors were seen throughout the human brain.

D_1 receptors were enriched in the caudate, putamen, olfactory tubercle, nucleus accumbens and medial globus pallidus as well as in the substantia nigra, pars compacta and reticulata (Figs. 1 and 2). Significant densities of D_1 receptors were also seen in other brain regions including the amygdala, the hippocampus and several cortical areas. However, the density of dopamine D_1 receptors in the latter areas was much lower than that found in the richly innervated dopaminergic areas. Dopamine D_1 receptors were not detected in the human cerebellum.

The highest densities of dopamine D_2 receptors in the human brain were also localized in parts of the basal ganglia including nucleus accumbens, nucleus caudatus, putamen, olfactory tubercle and in the lateral globus pallidus (Figs. 1 and 2). However, only low densities were seen in the medial globus pallidus.

Table 1. Densities of dopamine receptors in the adult human brain

Brain region	D$_1$ subtype[a]	D$_2$ subtype[b]
	[^{3}H]SCH 23390	[^{3}H] 205-502
Nucleus caudatus	267.5 ± 21	151.3 ± 11.6
Putamen	251.13 ± 21	163.7 ± 13.3
Nucleus accumbens	254.8 ± 33	191.4 ± 16
Globus pallidus, pars lateralis	27.7 ± 10	44 ± 9
Globus pallidus, pars medialis	102.8 ± 8	31 ± 6
Substantia nigra, pars compacta	131.9 ± 10	49 ± 12
Substantia nigra, pars reticulata	72 ± 8	31 ± 8
Olfactory tubercle	—	127 ± 14
Hippocampal formation		
CA1, stratum pyramidales	104 ± 25	41 ± 16
CA1, stratum lacunosum moleculare	—	31 ± 6
Dentate gyrus	14.5 ± 14.5	22 ± 4
Amygdala	—	14 ± 7
Cerebellum		
Granular layer	N.D.	2.1 ± 1
Molecular layer	8.9 ± 3.7	0.5 ± 0.5

[a] Data taken from Cortés et al. (submitted)
[b] Data taken from Camps et al. (submitted)
N.D. Non detectable binding
— Data not available
Values are in fmol/mg protein $\pm$ SEM (n = 1–15)

The substantia nigra presented relatively low densities of dopamine D$_2$ receptors, most of them localized to the pars compacta. Another area where dopamine D$_2$ receptors were seen was the hippocampus. In addition, the human pituitary contained also relatively high densities of dopamine D$_2$ receptors scattered throughout its anterior part, while no D$_1$ detectable binding was seen in this gland.

In general the distribution of dopamine receptors in the human brain is in good correlation with the distribution of endogenous dopamine, as measured biochemically and by immunohistochemistry (Bogerts, 1981; Scatton et al., 1982). Thus, the nucleus caudatus, putamen, nucleus accumbens and olfactory tubercle, which are known to be richly innervated by dopaminergic terminals, were also the nuclei exhibiting the highest densities of both dopamine D$_1$ and D$_2$ receptors. On the other hand, the substantia nigra, the area where the dopaminergic cell bodies are concentrated, presented moderate densities of dopamine receptors, although D$_1$ and D$_2$ subtypes were differentially located in this structure.

In addition, relatively high densities of the D$_2$ receptor subtype have been detected in the lateral part of the globus pallidus, an area known to contain

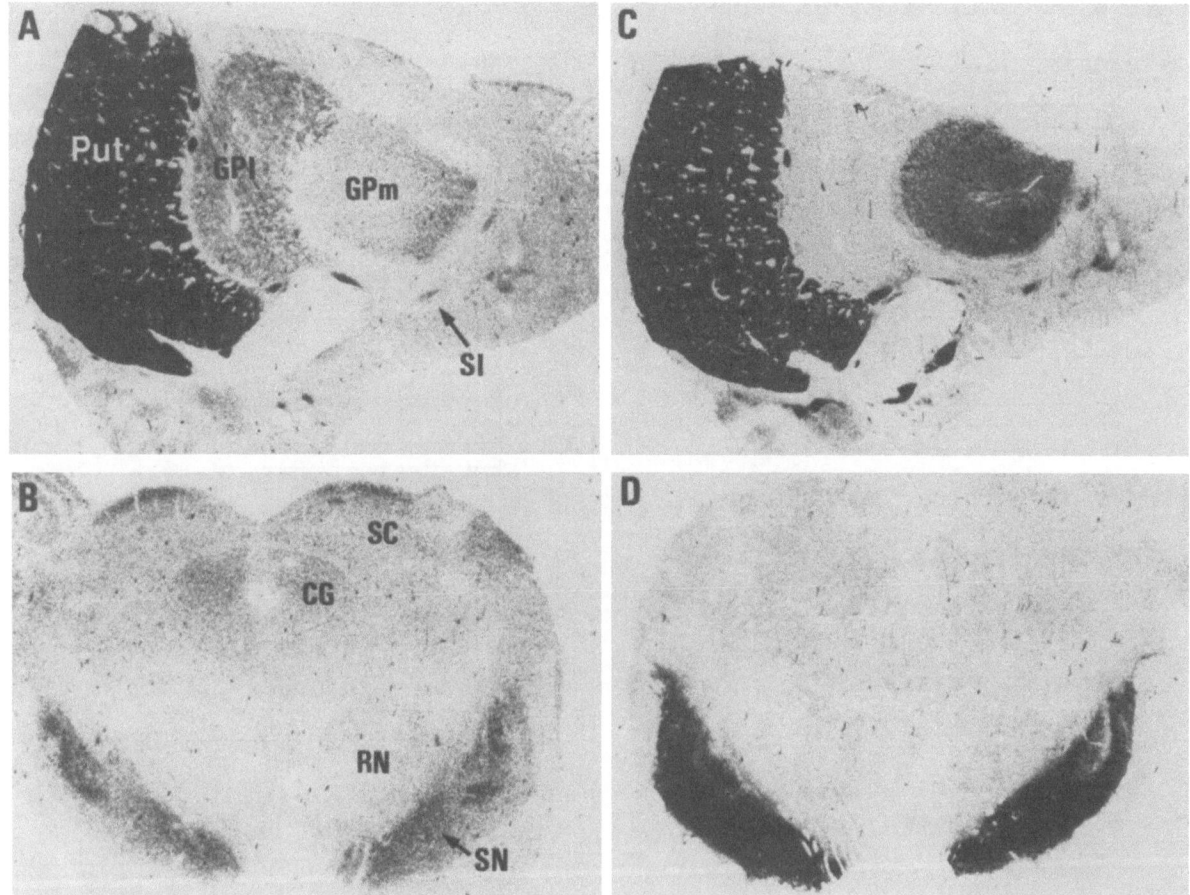

Fig. 1. Visualization of dopamine D_1 and D_2 receptors in the human brain. **A** and **B** Photomicrographs from autoradiograms of sections of a human brain showing the localization of dopamine D_2 receptors as labeled with [^{3}H]205-502. **A** High densities of D_2 sites in the putamen (*Put*), and lower densities in the lateral globus pallidus (*GPl*). Very low binding is seen in the medial globus pallidus (*GPm*) and substantia innominata (*SI*). **B** D_2 receptors in the midbrain. Moderate densities are observed in the substantia nigra (*SN*) and superior colliculus (*SC*). Very low densities are seen in the central grey (*CG*) and red nucleus (*RN*). Autoradiograms in **C** and **D** are from sections of the same patient shown in **A** and **B** illustrating the localization of D_1 receptors as labeled with [^{3}H]SCH 23390. D_1 receptors are enriched in the putamen, globus pallidus, pars medialis, and substantia nigra

significant amounts of endogenous dopamine. The medial part of this structure presents lower although still significant levels of endogenous dopamine which is in good correlation with the presence of D_1 receptors in this part of the globus pallidus.

Nevertheless, a good correlation between the dopaminergic innervation and the presence of dopamine receptors was not always seen. An example is the hippocampal formation where the dopaminergic innervation appears to be concentrated in the subiculum, while the dopamine receptors D_1 and D_2 were localized in dentate gyrus and CA 1, respectively. A lack of correlation between

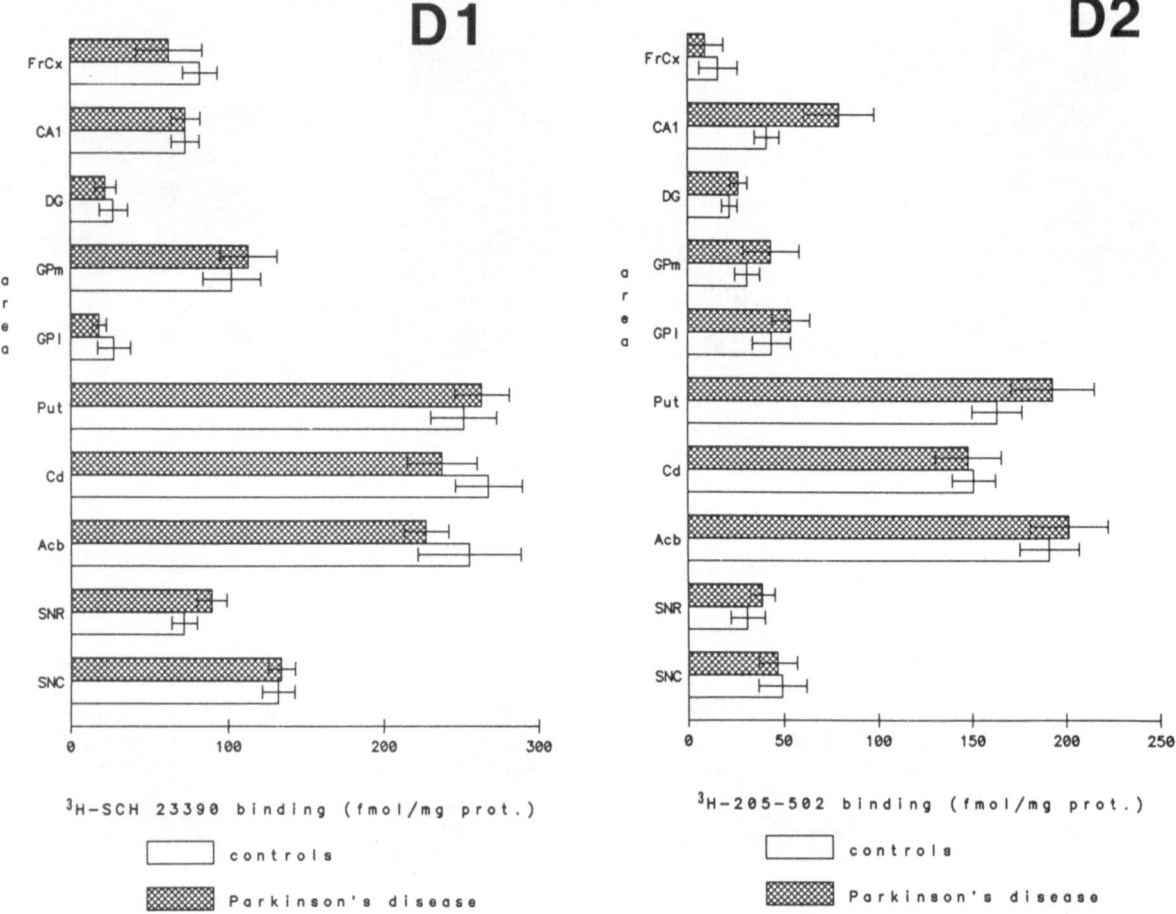

Fig. 2. Autoradiographic determination of D_1 and D_2 receptor densities in control and parkinsonian cases. D_1 receptors were labeled with 1 nM [^{3}H]SCH 23390 and D_2 receptors with 1 nM [^{3}H]205-502. Receptor densities were determined by microdensitometry as described in the text. Data are mean values obtained from 3 to 15 cases. No significant differences in the D_1 and D_2 receptor concentrations were found between parkinsonian cases and age matched controls. *FrCx* frontal cortex; *CA 1* CA 1 hippocampal field; *DG* dentate gyrus; *GPm* globus pallidus, pars medialis; *GPl* globus pallidus, pars lateralis; *Put* putamen; *Cd* nucleus caudatus; *Acb* nucleus accumbens; *SNR* substantia nigra, pars reticulata; *SNC* substantia nigra, pars compacta

innervation and receptor distribution has been observed for many other neurotransmitter systems and its signification extensively discussed (Kuhar, 1985).

Our results in the human brain are in good agreement with the distribution of dopamine receptors found in the laboratory animals (Gehlert et al., 1985; Dawson et al., 1986; Savasta et al., 1986 a; Charuchinda et al., 1987). In the rat the highest densities of dopamine receptors have also been found in the richly innervated areas such as the nucleus caudatus-putamen, globus pallidus, endopeduncular nucleus and substantia nigra pars compacta and reticulata, nucleus accumbens, olfactory tubercle and hippocampal formation. However, there are some differences in dopamine receptor distribution between rat and human brain. For example, we did not find dopamine D_1 and D_2 binding in the human cerebellum while there is a significant density of those receptors in the rat cerebellum.

In our studies we have analyzed the influence of postmortem delay, gender and age on the density of dopamine D_1 and D_2 receptors. In our population no effects of these factors were observed on D_2 receptor density. These results do not agree with recently reported age-dependent decreases in D_2 sites observed with positron emission techniques (Wong et al., 1984). In contrast, we found an age-related decrease in the density of D_1 receptors in the human brain. This effect was particularly marked in the substantia nigra, where a reduction of more than the 50% occurred between the first and the fourth decade of life. This decrease in D_1 receptors could be related to the reported decrease in dopaminergic cells occurring throughout life (McGeer et al., 1977).

We have also examined the densities of dopamine D_1 and D_2 receptors in several areas of the brain of patients dying with Parkinson's disease and compared with age-matched controls. We did not observed significant differences between the control and the parkinsonian population in any of the areas investigated, as shown in Fig. 2.

In addition, we did not see any change in the density of dopamine receptors in a case of progressive supranuclear palsy we have investigated. However, other groups have detected a marked decrease of dopamine receptors in several cases of progressive supranuclear palsy using binding assays (Ruberg et al., 1985) and positron emission tomography (Baron et al., 1985). These results of the investigations agree with the presence of dopamine receptors in the interneurons and/or neuronal cell bodies intrinsic to the caudate and putamen nuclei, as proposed from the results in animal experiments.

However, in our samples the absence of effects of Parkinson's disease on dopamine receptors in the basal ganglia and substantia nigra is surprising because it does not correlate with results from lesion experiments in the rat (Spano et al., 1977; Schwartz et al., 1978; Savasta et al., 1986 b). This suggests that there might be species differences in the localization of dopamine receptors in the human and rat brain. This clearly deserves further investigation.

References

Baron JB, Comar D, Zarifian E, Agid Y, Crouzel C, Loo H, Deniker P, Kellershohn C (1984) Dopaminergic receptor sites in human brain: positron emission tomography. Neurology 35: 16–24

Baron JC, Maziere B, Loch C, Sgouropoulos P, Bonnet AM, Agid Y (1985) Progressive supranuclear palsy; Loss of striatal dopamine receptors demonstrated in vivo by positron emission tomography. Lancet 1: 1163–1164

Bird ED, Iversen LL (1982) Human brain postmortem studies of neurotransmitter and related markers. In: Lajtha A (ed) Handbook of neurochemistry, vol 2. Plenum Press, pp 225–251

Bogerts B (1981) A brainstem atlas of catecholaminergic neurons in man, using melanin as a natural marker. J Comp Neurol 197: 63–80

234   J. M. Palacios et al.

Camps M, Cortés R, Zürcher B, Probst A, Palacios JM (1987) Dopamine receptors in human brain: autoradiographic distribution of D_2 sites. Submitted

Carlsson A, Lindquist J (1963) Effect of chlorpromazine and haloperidol on formation of 3-methoxytyramine and normetanephrine in mouse brain. Acta Pharmacol Toxicol 20: 140–144

Charuchinda C, Supavilai P, Karobath M, Palacios JM (1987) Dopamine D_2 receptors in rat brain: autoradiographic visualization using a high affinity selective agonist ligand. J Neurosci 7: 1352–1360

Cortés R, Probst A, Palacios JM (1984) Quantitative light microscopic autoradiographic localization of cholinergic muscarinic receptors in human brain: brainstem. Neuroscience 12: 1003–1026

Cortés R, Zürcher B, Pazos A, Probst A, Palacios JM (1987) Dopamine receptors in human brain: autoradiographic distribution of D_1 sites. Submitted

Crow TJ, Johnstone EC, Owen F (1979) Research on schizophrenia. Recent Adv Clin Psychiatry 3: 1–35

Dawson TM, Gehlert DR, McCabe RT, Barnett A, Wamsley JK (1986) D_1 dopamine receptors in the rat brain: a quantitative autoradiographic analysis. J Neurosci 6: 2352–2365

Farde L, Hall H, Ehrin E, Seduall G (1986) Quantitative analysis of D_2 dopamine receptor binding in the living human brain by PET. Science 231: 258–261

Friedmann AM, Dejesus OT, Woolverton WL, van Moffaert G, Goldberg LI, Prasad A, Barnett A, Dinerstein J (1985) Positron tomography of a radiobrominated analogue of the D_1/D_2 antagonist SCH 23390. Eur J Pharmacol 108: 327–328

Gehlert DR, Wamsley JK (1985) Dopamine receptors in the rat brain: quantitative autoradiographic localization using [^{3}H]sulpiride. Neurochem Int 7: 717–723

Hornykiewicz O (1972) Dopamine and its physiological significance in brain function. In: Bourne GH (ed) The structure and function of nervous systems, vol 6. Academic Press, New York, pp 367–415

Kebabian SW, Calne DB (1979) Multiple receptors for dopamine. Nature 277: 93–96

Kuhar MJ (1985) The mismatch problem in receptor mapping studies. Trends Neurosci 8: 190–191

Lee T, Seeman P, Rajput A, Parley IJ, Hornykiewicz O (1978) Receptor basis for dopaminergic supersensitivity in Parkinson's disease. Nature 273: 59–60

McGeer PL, McGeer EG, Suzuki JS (1977) Aging and extrapyramidal function. Arch Neurol 34: 33–35

Owen F, Crow TJ, Poulter M, Cross AJ, Longden A, Riley GJ (1978) Increased dopamine receptor sensitivity in schizophrenia. Lancet 2: 223–225

Palacios JM, Pazos A (1987) Visualization of dopamine receptors: a progress review. In: Creese I, Frazer C (eds) Structure and function of dopamine receptors. Alan R Liss, New York, in press

Palacios JM, Probst A, Cortés R (1986) Mapping receptors in the human brain. Trends Neurosci 9: 284–289

Pimoule C, Schoemaker H, Reynolds GP, Langer SZ (1985) [^{3}H]SCH 23390 labeled D_1 dopamine receptors are unchanged in schizophrenia and Parkinson's disease. Eur J Pharmacol 114: 235–237

Raisman R, Cash R, Ruberg M, Javoy-Agid F, Agid Y (1985) Binding of [^{3}H]SCH 23390 to D_1 receptors in the putamen of control and parkinsonian subjects. Eur J Pharmacol 113: 467–468

Rinne UK, Koskinen V, Lönnberg P (1980) Neurotransmitter receptor in parkinsonian brain. In: Rinne UK, Klinger M, Stamm G (eds) Parkinson's disease – current progress, problems, and management, vol 6. Elsevier North Holland Biochemical Press, New York, pp 93–107

Ruberg M, Javoy-Agid F, Hirsch E, Scatton B, Lheureux R, Hauw JJ, Doyckaerts C, Gray F, Morel-Maroger A, Rascol A, Serdaru M, Agid I (1985) Dopaminergic and cholinergic lesions in progressive supranuclear palsy. Ann Neurol 18: 523–529

Savasta M, Dubois A, Scatton B (1986a) Autoradiographic localization of D_1 dopamine receptors in the rat brain using [^{3}H]SCH 23390. Brain Res 375: 291–301

Savasta M, Dubois A, Benavides J, Scatton B (1986b) Different neuronal location of [^{3}H]SCH 23390 binding sites in pars reticulata and pars compacta of the substantia nigra in the rat. Neurosci Lett 72: 265–271

Scatton B, d'Ambrosio A, Javoy-Agid F, Agid Y, Bischoff S, Simon H, le Moal M (1982) Evidence for the existence of a dopaminergic innervation of the rat and human hippocampal formation. In: Koshaka M, Shohmori T, Tsukada Y, Woodruff GN (eds) Advances in dopamine research. Pergamon Press, Oxford, pp 337–382

Schwartz R, Creese I, Coyle JT, Snyder SH (1978) Dopamine receptors localized on cerebral cortical afferents to rat corpus striatum. Nature 271: 766–768

Spano PF, Govoni S, Trabucchi M (1977a) Studies on the pharmacological properties of dopamine receptors in various areas of the central nervous system. Adv Biochem Psychopharmacol 16: 307–310

Spano PF, Trabucchi M, di Chiara G (1977b) Localization of nigral dopamine-sensitive adenylate cyclase on neurons originating from the corpus striatum. Science 196: 1343–1345

Stoof JC, Kebabian JW (1984) Two dopamine receptors: biochemistry, physiology and pharmacology. Life Sci 35: 2281–2296

Unnerstall JR, Niehoff DL, Kuhar MJ, Palacios JM (1982) Quantitative receptor autoradiography using [^{3}H]Ultrofilm: application to multiple benzodiazepine receptors. J Neurosci Methods 6: 59–73

Wagner HN, Burns HD, Dannals RF, Wong DF, Langström B, Duelfer T, Frost JJ, Ravert HT, Links JM, Rosenbloom SB, Lukas SE, Kramer AV, Kuhar MJ (1983) Imaging dopamine receptors in the human brain by positron emission tomography. Science 221: 1264–1266

Wong DF, Wagner HN, Dannals RF, Links JM, Frost JJ, Ravert HT, Wilson AA, Rosenbaum AE, Gjedde A, Douglas KH, Petronis JD, Folstein MF, Toung JKT, Burns HD, Kuhar MJ (1984) Effects of age on dopamine and serotonin receptors measured by positron emission tomography in the living human brain. Science 226: 1393–1396

Authors' address: J. M. Palacios, Preclinical Research, Sandoz Ltd., CH-4002 Basle, Switzerland.

J Neural Transm (1988) [Suppl] 27: 237–240

Discussion

Biochemical studies in parkinsonism

Melamed: About the results of the ascorbic acid and glutathione in MPTP-treated animals, would you care to try and speculate why you see those changes in animals treated with MPTP?

Riederer: I would guess that if we believe in the radical theory of MPTP toxicity, that these radicals react first with glutathione. That might be the reason why we have an increase in GSSG and that the concentration is high enough in mice at least to overcome for a long time the severe toxicity that we would see in other animals, and also in man.

Leenders: Do you think that the native dopamine content of the striatum actually reflects dopaminergic neuron cell loss in the striatum? Connected so that, in early parkinsonian patients, what level of loss do you think have they compared to the global loss? Lastly, what do you think is the monoamine oxidase B-level in parkinsonian patients?

Riederer: I start with the last question. The monoamine oxidase type B-activity in Parkinson's disease is normal in most of the brain areas except maybe for the substantia nigra in malignant cases; there you have the tendency at least for an increase. The second question was early Parkinson. We have done a study in 1976, showing a dopamine decline in the caudate nucleus with time and with duration of the disease. We collected a lot of samples and measured them. I would guess when you start with Parkinson's disease and the patient is dying let's say one year after start of clinical Parkinson's disease then the loss is already, at least in our experience, around 70%. That is hard data. We do not know what is before; we know that in the end stage the loss in the akinetic crisis is about 90% or more, as I have shown. But there is a non-linear decline.

Klawans: The iron story becomes more complicated. If you give neuroleptics chronically to experimental animals in the animal models of tardive dyskinesia you certainly raise brain iron. In our hands giving levodopa chronically leads to an increase in brain manganese and an increase in brain iron. That would raise the question as whether we may be doing more harm than good with a variety of things that we do that can chelate things and bring substances into the brain?

Marsden: The iron story, I think, is very complicated. To add to the complication, there is this debate amongst the Magnetic Resonance Imaging doctors claiming that they can see changes in the substantia nigra on MRI suggesting to them increase deposition of iron in the substantia nigra in Parkinson's disease. But there is another group who say that they can't see this. The point of the argument is not so much which side is right or wrong, but the fact that measurements of iron content, the sort that Peter Riederer presented very nicely, actually doesn't tell you in detail whether the iron is accessible to drive these free radical generation mechanisms. In fact, it is very difficult, I believe, to measure the free iron content that may be actually generating free radicals. It is a very complicated business.

Gessa: The difference between the toxicity to MPP$^+$ in rats and marmosets, is there any possibility that this is due to differences in the uptake capability of the substantia nigra of the two species? I have evidence that in rats the substantia nigra does not take up dopamine, that means dendrites are unable to pick up dopamine. I wonder whether the monkey has the same impairment. You have shown data on mazindol preventing MPTP-toxicity. This is maybe due to the fact that it is preventing uptake.

Marsden: To my knowledge, I don't think anybody has looked at the dopamine uptake system in the primate, nor at the dopamine uptake system in nerve terminals as against substantia nigra dendrites. If one takes the rat data, then there is an interesting gap in the MPTP-story that has yet to be filled in. If the main dopamine uptake system is in nerve terminals in the striatum, then you have got to get the MPP$^+$ back down the axon into the perikaryon to bind to the neuromelanin. That does introduce an interesting concept, that MPTP via MPP$^+$ might actually have two methods of neurotoxicity. One is producing toxicity in the nerve terminals where it is taken up in the striatum; that is mainly what you see in the rodent, with a lot of recovery from terminal-sprouting. The second form of toxicity is after retrograde axonal transport back to neuromelanin in the cell body which produces irreversible cell damage. But that is a step, I think, in the whole story that is not certain. One observation: Although you can prevent MPTP-toxicity very well with dopamine reuptake blockers in rodents, there is considerable argument and debate about how efficient drugs like mazindol are in preventing toxicity in primates.

Parkes: Does the firing pattern, if you stick the needle in the right place, always correspond to that pharmacological behavioural definition or can it be different? Do the electrical findings correspond to the pharmacological findings?

Mereu: You were asking if there is an electrical response which corresponds to the pharmacological classification. I mentioned at the end of the presentation that, for instance, some of the ergot derivatives are considered to be without intrinsic activity. So we can consider these compounds such as lisuride as a pure antagonist. In fact, these compounds when administered activate the cell more than a mixed agonist/antagonist. On the other hand, if we administer for instance N-propylnorapomorphine, which is an absolutely pure agonist, the effect of this compound is only to inhibit at very low doses.

Parkes: Can you help me on firing rates? I seem to remember some old literature suggesting that the nigrostriatum was one of the few areas of the brain in whom the spontaneous firing rate of dopamine neurons across sleep and waking was fairly constant and didn't vary much, despite all we have heard today about circadian variation in parkinsonian symptoms. Is that correct?

Mereu: As far as I know, yes that is correct. It is not clear what is a spontaneous activity, waking/sleeping cycle, and what is a pharmacologically induced sleep by an agonist that is controlled by nigrostriata or in any case dopaminergic neurons.

PET scanner studies and receptor mapping of the human brain

Chase: I was wondering whether when you see fluorodopa you can actually see the other portions of the dopamine system, the mesocortical system and so forth. Obviously the nigrostriatal system is very well outlined, but I couldn't notice in your pictures the other portions of the dopamine pathways.

Leenders: The striatum is so dominant that everything lower is scaled out of the colour scales. The second thing is, our machine was not very sensitive and I had to work with very low columns. The new system hopefully will clear that up a bit and might improve it.

Chase: The other question that I was curious about is that an hour after you give i.v. fluorodopa there is no apparent difference between normal controls and your Parkinson patients. Is that correct? The differences only become statistically significant after maybe two hours or so.

Leenders: Apparently. After one hour, enough dopa is converted to dopamine to have its clinical effect but may not tie in with the eventual maximum capacity to store dopamine. From these data I don't think you can relate those two things directly. I think the data are too crude, they overlap too much to make a specific comment about the relationship.

Rinne: You demonstrated a mild decline of ^{3}H-spiperone-binding in levodopa-treated advanced cases. And I realize that it fits very well with our recent post-mortem data.

Marsden: That is fascinating data, but very worrying data. The discrepancy between the localization of D1 and D2 apparent receptors is intriguing. Can you tell us a bit about how you define specificity of your due ligands?

Palacios: We do all the standard procedures. The binding is saturable of high affinity, and reversible with the proper pharmacology. Proper pharmacology means that it is displaced stereospecifically by compounds with the right level of activity.

Marsden: I have no doubt you have done the proper pharmacology, but can you be even more specific. Is there no non-specific binding of either of your ligands?

Palacios: There is a non-specific binding. In fact, I illustrated a non-specific binding for the dopamine D2 ligand in one of the slides from the human hippocampus. But the non-specific binding is relatively homogeneous, so in this sense I am not worried about non-specific binding.

Marsden: The last question then: In the area where there is the greatest discrepancy, i.e. in the nigra, where you have the D2 ligand and the D1 ligand labelling very different areas, all that is specific binding?

Palacios: Yes. The differences are reproduced in at least five species.

Gessa: The presence of dopamine receptors in areas like the cerebellum where no dopamine neurons are present, do you suggest that dopamine is released by norepinephrine neurons?

Palacios: I have no idea, but this is not an uncommon result. There are receptors for many neurotransmitters in areas where the presence of the endogenous neurotransmitters has not been reported. The question is, who is wrong. This is what is called a mismatch problem. What I can tell you is that these receptors appear to be functional, because they are linked to a response. If you take a cat cerebellum and look for a dopamine sensitive adenylate cyclase, you will be able to find this cyclase with the proper pharmacology.

Present state and future developments

J Neural Transm (1988) [Suppl] 27: 243–247

Discussion

Different drug releasing systems

Obeso: Do you have any data regarding changes in the temperature of the skin or the blood flow or some other factors which could become important factors modifying the capacity of transdermal absorption?

Stahl: It is true that skin blood flow is variable and can be important for certain drugs. This has been studied in the case of clonidine and nitroglycerin. It turns out that within the vicissitudes of the ups and downs of blood flow to the skin they are not rate limiting, so that even at the lower end of blood flow the rate control is still in the patch for those systems. That may not be the case particularly for subcutaneous administration, nor for those drugs whose absorption kinetics are such that they will be slower than blood flow. So it is a consideration, but in practice it has not been an issue for the currently marketed transdermal products.

Obeso: I just wanted to press you on what you are thinking of a dopamine agonist we are going to apply to patients with on-off. I don't know really, but I suppose that a possible problem could be that when the patient goes off, the skin gets cold, therefore you are getting to a negative vicious circle, the more off they are, the less blood is flowing through the skin, the less drug goes in.

Stahl: Skin temperature is not that variable on surfaces, I have read. It again is something we should be aware of, but if a dopamine agonist or a narcotic analgesic work like the drugs for the cardiovascular system it may turn out not to matter.

Marsden: Can you expand on the business of absorption of peptides through the cribriform plate. What parts of the brain do they get into via the nose?

Stahl: Very poorly investigated. I wish some people here and elsewhere would take that up. What is known is that a number of peptides have been extensively administered in humans through the nose. But as far as I know no one has done any tracer studies. There is not any radioactivity study to show where they go, whether they get into the olfactory nerves, whether they just leak into holes in the plate. It has been speculation that it is an arterial delivery through the nose. This is one of the reasons why cocaine is thought to be such a potent dopamine agonist. It is essentially like giving an intraarterial injection or intracarotid injection. Drug delivery to the brain is most efficient when it is arterial, it is thought.

Infusion technology

Parkes: Can we take a straw poll of those using or thinking of using sub-cutaneous lisuride infusion and see really those in favour of constant 24-h delivery or those in favour of entire delivery during the day with none during the night? In practical terms, if we are doing this in 10 years' time, what will we be doing?

Chase: My prediction is that we will use continuous stimulation, because after all that is the physiologic way that the central nervous system works. The dopamine system fires off at about the same rate all the time. If we want to restore normal function, then we have to let the postsynaptic dopamine receptors see about the same amount of dopamine or dopamine agonists all the time. In terms of our own experience with i.v. infusions, we have seen no difference between exercise and no exercise in terms of dopamine requirement. We do have some data on this, and I cannot demonstrate that exercise increases the requirement for levodopa in a Parkinson patient.

Parkes: In a land of darkness for most of the year, i.e. Finnland, Urpo Rinne what are you going to do?

Rinne: I think that it is good to have a continuous stimulation, but you have to consider that during daytime you need more than during night. I prefer to give more during daytime and about half or one quarter during night.

Parkes: Can we take a vote? Those who are going to give 24-h constant infusion hands up. Those who are going to give variable infusion rates? The variables have a clear majority, but they of course may not be right.

If we are going to use a drug delivery system, what is going to be our target to look at? We are left with everything from Lewy bodies through different receptors, through CCK-8, through something inside the cell, through indeed just trying to get the drug non-specifically to a D 1 or D 2 receptor. What is going to be your target, Steven?

Stahl: I am afraid that the distribution of receptor selectivity is going to work against us. I think the best we can do in the 20th century is to target enzymes and receptors. In those cases where the Good Lord made a CCK receptor in one part of the brain different than the other, then we'll be all right. In those cases where he didn't, I don't think we are going to have it this century.

Parkes: Fuxe suggested that CCK-8 was present in the mesolimbic system and CCK-37 in the nigrostriatal system, and this might provide some way of selectively affecting these two different systems. But this is theory at present, I think, rather than fact. Now, in the next 10 years doubtless many embryonic primate catecholamine-producing cells will be implanted into humans as well as MPTP-monkeys.

Melamed: I doubt that this will offer the ultimate solution to the problem, because maybe what is needed is all the wiring between the nigra and the striatum. When you implant embryonic cells into the striatum this will not be enough. I'd like to comment on what Tom Chase said. The question is whether when we are flooding the striatum and other central dopaminergic terminal fields with levodopa or with lisuride we are really mimicking the physiological condition. We are not, I think. We do not know what is the requirement of dopamine at the synaptic clefts. We do not know the meaning of the tonic release of dopamine. We do not know when and why and how dopamine neurons fire and all the complex wiring systems, the feedback loops.

Parkes: David Marsden, have you got any more information on embryonic transplants in MPTP-monkeys?

Marsden: Yes, Steve Dunnett has transplanted two of our marmosets and confirms what the Americans have published, that you can get fetal marmoset nigras to grow in a MPTP-treated marmoset and restore some degree of function. I would take a different view than Eldad Melamed. I think, one of the most remarkable things Steve Dunnett's work in collaboration with Anders Björklund has been to show that you can restore not just crude functions with dopamine transplants but very subtle deficits in at least rodents. Even more remarkable, I think, is the restoration of function with acetylcholine transplants. Dopamine is a curious substance and not perhaps a point to point neurotransmitter. But acetylcholine one would have thought is a plus/minus classical neurophysiological transmitter. The fact that you can restore function suggests that perhaps the wiring does get itself sorted out naturally.

Parkes: Two not really opposing views, they are suggesting that we will be doing transplants in parkinsonism, but perhaps in other diseases as well.

Stahl: I think that it may be that one needs to kick-start the system with a bolus and then to maintain it. The substantia nigra which is left, the last 5 or 10%, all it's got to do is make the difference between when you need it on demand or not. Maybe even a subthreshold amount of constant stimulation will allow the remaining nigra to work. I think you could do that through various pumps.

Klawans: I would also conceive that you could make a pill that would drop off some for an immediate kick-off, first thing in the morning, and then give you a slow steady delivery for the next 12–14 hours. That shouldn't be very difficult. So that it may well be a combination of a variety of release mechanisms within a single pill at some time.

Parkes: The next question is, of course, what are we going to put in this pill. Are we going to have a D 1 and a D 2 agonist mixed with a monoamine A and a monoamine B inhibitor? Are we going to have a D 2 antagonist, a D 1 agonist or vice versa? What is this pill going to be? What do we really think, in 1996, will be in this morning pill taken by patients with Parkinson's disease?

Horowski: It should be a highly effective and a little bit intelligent D 2 agonist that has got 100% bioavailability and a long half-life.

Klawans: I'll take the devil's advocate point of view from the other side. I think it will end up being levodopa. Levodopa infusions seem to work in our chronic patients that are having problems. Let's find a way of infusing it through a pill and giving us constant delivery into the gut. And even though the medication itself is not patentable, the delivery system will be, so it still is worth investigating.

Parkes: I'll end up just again by asking the basic scientists in the audience, why they think we have got this switch in the brain delivering a fairly constant rate of soup to just a small area of the brain? We have said that really the delivery of this or the firing rate of these neurons is remarkably stable; it doesn't depend all that much on exercise or sleep or wakefulness. What on earth is it doing? In models of other dopamine systems, for example the growth hormone system, ones response to levodopa is all or nothing, ones response to apomorphine is all or nothing. If we look at the prolactin system, it is different. Lisuride is binding irreversibly to the prolactin receptor and turning the system off for a prolonged period. If you look at the carotid body, hypoxic eucapnic drive which is dopamine-mediated, the response is very different. I am saying that if I breath a low concentration of oxygen, the firing rate of my carotid body mediated via dopamine systems will alter my central ventilatory response and I will depress my ventilation. This pattern is linear over a certain curve. Here we have 3 or 4 totally different dopamine control switches. We have got an all or none switch growth hormone, we have got a constant turn-off prolactin switch, we have got perhaps a constant go-switch in the nigrostriatum and we have got a graded response in 3 or 4 breaths to hypoxia in the carotid body.

New pharmacological approaches and future developments

Horowski: I want to ask Prof. Przuntek to shortly comment on his experience with lisuride in malignant hyperthermia of the neuroleptic type.

Przuntek: We have treated three patients with a neuroleptic syndrome, and two of these patients had a classical malignant neuroleptic syndrome. We started with low levels because we didn't know how much they tolerated and we had to increase the dosage up to between 4 and 6 mg of lisuride per day. The third had another neuroleptic syndrome; it was more malignant, but there was no hyperthermia or tachycardia, and no increase of CPK. This girl was injected with haloperidol; she was unable to cough and she developed pneumonia and atelectasis of the lungs. Therefore, the doctors from the internal medicine asked me to treat this patient with lisuride, too. This girl improved within 4 days. When we started the therapy, the girl was unable to cough, and after 4 days' treatment with 6 mg lisuride per day she was able to eat, to speak and to cough again.

Stahl: You have an interesting family of full and partial agonists and antagonists. We have learned, I think, from the beta and cholinergic systems how differentially these drugs behave in innervated versus denervated systems. In addition to controlling rates which go up and down, isn't it possible that we should have a family of agonists in various stages of denervation? What I am thinking is that in the denervated nigrostriatal pathway a partial agonist is very full when there is no endogenous dopamine competing, but in the intact mesolimbic system it would not be very full because there is dopamine and you wouldn't get psychoses. Contrarywise, a partial antagonist could block the supersensitivity of an overactive mesolimbic and be a better neuroleptic, whereas by not doing the same thing in the striatum which is intact in schizophrenia wouldn't cause acute extrapyramidal or chronic tardive effects. Have you thought of this?

Horowski: Yes, we are thinking about this along the same lines. It may be simplistic but we have to test this interesting and important question. We are studying in schizophrenics now bromerguride which may be too much antagonist, and terguride which may be too much agonist, so that we can find finally something in between. This would be of course very important to study, the problem being that studies in schizophrenia are rather difficult. We pharmacologists develop so many good new compounds, but clinicians sometimes have very big difficulties in giving us exact measures of how much agonistic or antagonistic they are.

Gessa: Is there any study on the use of these dopaminomimetic agents, partial agonists and so on, on hyperkinetic children? The reason why I am asking is, as you know amphetamine is used and paradoxically causes a sort of sedative effect. This might be due to the fact that in these conditions the prevailing effect of amphetamine is to decrease the hyperactivity of the dopaminergic system, at least this may be a possible explanation.

Horowski: There has been one study indeed, more than 10 years ago, with lisuride in this indication in Czechoslovakia, and they have reported beneficial effects. However, we didn't follow up this for various reasons.

J Neural Transm (1988) [Suppl] 27: 249–252

Continuous dopaminergic stimulation: state of the art and outlook

R. Horowski[1], **C. D. Marsden**[2], and **J. A. Obeso**[3]

[1] Clinical Research, Schering AG, Berlin/Bergkamen, [2] University Department of Clinical Neurology, Institute of Neurology, London, U.K., [3] Clinica Universitaria, Departmento de Neurologia, Universidad de Navarra, Pamplona, Spain

Therapy with L-Dopa was a major break-through in the treatment of Parkinson's disease. It was an important precedent, and a stimulus for research in this and other fields of neurology. Initial side effects associated with this drug were overcome by refining the treatment schedule and by additional drugs inhibiting the peripheral conversion of L-Dopa to dopamine.

On long-term treatment with L-Dopa, however, fluctuations in motor performance occur in a majority of cases. These respond only in part to an increased and fractionated intake of L-Dopa. Results obtained with short-term L-Dopa infusions have shown that pharmacokinetic factors play a role in this situation (Birkmayer and Hornykiewicz, 1961; Shoulson et al., 1975; Quinn et al., 1982). For galenical reasons, intravenous L-Dopa treatment can be given only for a short period, so it became necessary to devise new ways of achieving an easily controllable and constant dopaminergic stimulation, both for further research and treatment.

At this meeting held in Alicante, a number of clinical groups have confirmed and extended the original findings by J. A. Obeso and his colleagues (1986). These authors found that continuous s.c. infusion of lisuride, a watersoluble dopaminergic 8-α-aminoergoline with dopaminergic properties (Horowski, 1983) which can be injected or infused (Dorow et al., 1980), can improve— sometimes quite considerably—motor function in severely disabled fluctuating parkinsonian patients. The concurrent use of the peripheral dopamine antagonist domperidone attenuates or prevents side effects related to the stimulation of "peripheral" dopamine receptors, including the chemoreceptor trigger zone and some areas of the hypothalamus outside the blood-brain barrier.

In addition, as predicted from the clinical pharmacology of dopaminergic drugs (Horowski, 1986 a), some side effects seem to depend less on a given drug level in the blood but more on the rapid increases and oscillations of dopaminergic stimulation which are inevitable with oral treatment. Furthermore, some of these dopaminergic effects such as nausea, emesis, orthostatic hypotension and increased GH levels rapidly undergo tolerance, whilst the antiparkinson effects persist in the same patients and other effects even can increase with time. This means that patients treated with oral dopaminergic therapies undergo a form of "dissociated functional tolerance/supersensitivity development" on repeated treatment.

The results obtained with chronic s.c. lisuride infusion have shown that motor function can be restored even in severely disabled patients, whilst peripheral side effects are of little importance in spite of the strong "dopaminergic load". However, central side effects such as dyskinesias and mental changes may become a crucial and limiting factor in such patients, who have a long-lasting history of disease and L-Dopa treatment. In particular, the problem of gross psychic disturbances which have been observed in a significant number of patients, has been discussed at length at this meeting and subsequently (Critchley et al., 1986; Bittkau and Przuntek, 1986; Ruggieri et al., 1986; Horowski, 1986 b). It appears that relevant factors conditioning the appearance of psychic disturbance are the severity and duration of the disease, length of L-DOPA therapy, age, dementia, concomitant treatment with anticholinergics (and possibly even benzodiazepines) and above all a previous history of psychiatric complications induced by L-DOPA or other dopaminergic therapies. It is hoped that more stringent patient selection (excluding those with a history of psychosis), starting the infusion therapy in the earlier stages of fluctuations, an intermittant form of treatment (12 hours per day), avoidance of anticholinergic drugs, and close monitoring of patients for early warning symptoms will enable clinicians to avoid these side effects which can be very distressing for patients and their relatives.

In the meantime, the beneficial effects of continuous dopaminergic stimulation on motor function in severely disabled patients has been confirmed independently by one group present at this meeting who used apomorphine instead of lisuride in a s.c. infusion over 12 hours per day (Stibe et al., 1987). Controlled studies are necessary to decide whether this form of therapy differs as regards efficacy or side effects; it has to be mentioned that early studies with parenteral apomorphine not only demonstrated antiparkinson efficacy (Schwab et al., 1951; Struppler, 1953), but also reported psychiatric side effects (Strian et al., 1972). In fact, Obeso et al. (1987) have reported transient psychiatric side effects during apomorphine infusion in patients who previously have had similar complications with subcutaneous lisuride infusion. Clearly, great progress could be achieved if central side effects could be controlled in the same way as the peripheral problems. Concomitant treatment with antipsychotic drugs not interfering with dopaminergic motor systems, such as clozapine (Scholz and Dichgans, 1985) or, possibly, selective 5-HT-antagonists, need to be investigated in this respect, as well as more selective dopamine agonists or partial agonists.

In addition to new drugs, other new ways of delivering continuous dopaminergic effects have to be developed. Intraduodenal application, as reported by Duvoisin at this meeting, points to the importance of local enteral effects as well as hepatic first-pass metabolism (Burns and Calne, 1983), which in future possibly could be overcome by transdermal or other new delivery systems.

It is not yet clear whether direct continuous dopaminergic stimulation delivered by transplanted cells will become of clinical importance. Diagnostic improvements and new concepts in basic knowledge may give us further information on this important issue.

Finally, the clinical results discussed in this volume may not only be a basis for further improvements in our knowledge and therapeutic strategies in Parkinsonism; they point, more generally speaking, to the so far neglected importance of different ways of stimulating neurological or other systems, e.g. discontinuous, oscillatory effects caused by frequent oral application vs. continuous stimulation as described here with the lisuride s.c. infusion. It is known from endocrinology that pulsatile stimulation may even have opposite effects compared to continuous stimulation (e.g. in the case of LHRH). Similar concepts have to be discussed and investigated in neurological disorders. In this respect, this multidisciplinary meeting and its publication may offer new ideas and concepts for therapy in general, in addition to its potential application in the treatment of the complications of Parkinson's disease.

References

Birkmayer W, Hornykiewicz O (1961) Der L-3,4-Dihydroxyphenylalanin (= Dopa)-Effekt bei der Parkinson-Akinese. Wien Klin Wochenschr 73: 787–789

Bittkau S, Przuntek H (1986) Psychosis and the lisuride pump. Lancet 2: 349 (letter)

Burns RS, Calne DB (1983) Disposition of dopaminergic ergot compounds following oral administration. In: Calne DB, Horowski R, McDonald RJ, Wuttke W (eds) Lisuride and other dopamine agonists. Raven Press, New York, pp 153–160

Critchley P, Grandas Perez F, Quinn N, Coleman P, Parkes D, Marsden CD (1986) Psychosis and the lisuride pump. Lancet 2: 349 (letter)

Dorow R, Gräf KJ, Nieuweboer B, Horowski R (1980) Intravenous lisuride: a new tool for testing responsiveness to dopamine agonists and neuroendocrine function. Acta Endocrinol [Suppl] (Copenh) 234: 9

Horowski R (1983) Pharmacological effects of lisuride and their potential role in further research. In: Calne DB, Horowski R, McDonald RJ, Wuttke W (eds) Lisuride and other dopamine agonists. Raven Press, New York, pp 127–139

Horowski R (1986 a) Pharmacology of the dopamine agonist lisuride and its potential in the treatment of parkinsonism. In: van Manen J, Rinne UK (eds) Lisuride: a new dopamine agonist and Parkinson's disease. Excerpta Medica, Amsterdam, pp 24–35

Horowski R (1986 b) Psychiatric side-effects of high-dose lisuride therapy in Parkinsonism. Lancet 2: 510 (letter)

Obeso JA, Luquin MR, Martinez-Lage JM (1986) Lisuride infusion pump: a device for the treatment of motor fluctuations in Parkinson's disease. Lancet 1: 467–470

Obeso JA, Grandas, F, Vaamonde J, Luquin MR, Martinez-Lage JM (1987) Apomorphine infusion for motor fluctuations in Parkinson's disease. Lancet 1: 1376–1377

Quinn NP, Marsden CD, Parkes JD (1982) Complicated response fluctuations in Parkinson's disease: Response to intravenous infusion of levodopa. Lancet 2: 412–415

Ruggieri S, Stocchi F, Agnoli A (1986) Lisuride infusion pump for Parkinson's disease. Lancet 2: 348–349 (letter)

Scholz E, Dichgans J (1985) Treatment of drug-induced exogenous psychosis in Parkinsonism with Clozapine and Fluperlapine. Eur Arch Psychiatr Neurol Sci 235: 60–64

Schwab RS, Amadori LV, Lettoni JY (1951) Apomorphine in Parkinson's disease. Trans Am Neurol Assoc 76: 251

Shoulson I, Glaubiger GA, Chase TN (1975) On-off response: clinical and biochemical correlations during oral and intravenous levodopa administration in parkinsonian patients. Neurology 25: 1144–1148

Stibe C, Lees A, Stern G (1987) Subcutaneous infusion of apomorphine and lisuride in the treatment of Parkinsonian on-off-fluctuations. Lancet 1: 871

Strian F, Micheler E, Benkert O (1972) Tremor inhibition in Parkinson syndrome after apomorphine administration under L-Dopa and decarboxylase—inhibitor basic therapy. Pharmacopsychiatry 5: 198–205

Struppler A (1953) Die therapeutische Beeinflußbarkeit des Tremors. Med Monatsschr 73: 157–160

Authors' address: Dr. R. Horowski, Clinical Research, Schering Berlin/Bergkamen, Müllerstrasse 172, D-1000 Berlin 65.

Subject index